Landscape as Dialogue

Landscape as Dialogue redefines the process of understanding landscapes for students and practitioners so they can create more integrated, healthy places. Traditional site analysis sees the landscape as a series of components, evaluated individually, before being put back together. This perpetuates existing social hierarchies, maintains the need for high energy inputs and trumpets iconic designs that contribute to gentrification. This book examines the process of landscape dialogue as a natural give and take with the environment, drawing on diverse and challenging writings from design, geography, philosophy and ecological sciences to probe the relationship between humans and landscape. Each chapter begins with a discussion of a theoretical approach to landscape dialogue, such as perception, information or critique, before offering a series of practical steps and representation techniques that designers can use in understanding the landscape. Detailed illustrated case studies from around the world, including Hawaii, the American Southwest, Japan, China, Mexico, Turkey and Peru, explore the book's lessons in practice. This must-read book offers a radical alternative to conventional analytical approaches, inspiring designers to fully engage in the landscape to ultimately generate ecologically considered places.

Cory Parker is a landscape architect and geographer who teaches at the University of California, Davis.

Landscape as Dialogue

A New Approach to Site Analysis and Design

Cory Parker

Routledge
Taylor & Francis Group

LONDON AND NEW YORK

Designed cover: Cory Parker.

First published 2025
by Routledge
4 Park Square, Milton Park, Abingdon, Oxon OX14 4RN

and by Routledge
605 Third Avenue, New York, NY 10158

Routledge is an imprint of the Taylor & Francis Group, an informa business

British Library Cataloguing-in-Publication Data
A catalogue record for this book is available from the British Library

ISBN: 978-0-367-74655-1 (hbk)
ISBN: 978-0-367-74653-7 (pbk)
ISBN: 978-1-003-15894-3 (ebk)

DOI: 10.4324/9781003158943

Typeset in Minion Pro and Avenir
by KnowledgeWorks Global Ltd.

Contents

About the author

Cory Parker researches, teaches, writes on and lives within the landscape. He is an adjunct professor at the University of California, Davis in Human Ecology. He completed a Spatial Justice Fellowship as a visiting professor at the University of Oregon in the Design School. He received his PhD in geography from U.C. Davis focusing on the homeless experience of movement in several California cities. Before the university, Cory practiced landscape architecture of public infrastructure for 20 years, primarily working on large transportation and park projects in the United States at Jones & Jones Architects and Landscape Architects, Seattle, Washington.

Introduction

We do not gather information from the landscape, we converse with it.

The *kölea*, a small, speckled bird with stilt-like legs, summers on the Alaskan tundra, ground foraging for invertebrates, berries and seeds. After breeding in the fall, it flies south to islands in the Pacific to winter. Also called the Pacific golden plover, many *kölea* land their emaciated bodies on the Hawaiian islands. *Kölea* run-stop-run across the lawns of golf courses searching for food to fatten up. As Trask invokes, native Hawaiian islanders call tourists from the West *haole ki kölea* in honor of this seasonal immigrant. Islanders decry the extractive lifestyle of the

0.1 A breeding male Pacific golden plover.

DOI: 10.4324/9781003158943-1

kōlea, particularly developers and tourists who come to the island, stay a short time and leave plump and happy.

* * *

I clamber up the creek embankment and step over the guardrail along the shoulder of Kamehameha Highway on the windward side of Oahu. Steep hillsides covered with the ethereal but invasive *Albizia* trees surround us, framing scenes of low-slung houses and small-scale agriculture in the middle ground. In the foreground, narrow-leaf carpet grass covers the swale flowing over the highway embankment. On the far side of the road, a gas station hosts a couple of trucks parked right up against the wall of a mini-mart. Beyond that, an enormous banyan tree offers shade to a couple of older men sitting in cheap plastic chairs.

I signal to the other landscape architects that I have finished taking notes and we return to the car perched off the edge of the pavement. The four of us – two designers from Honolulu and two of us from Seattle – are evaluating the sustainability and cultural significance of the State of Hawaii's road network for the Department of Transportation. One landscape architect from Honolulu is driving the two of us around the island, while the other films the road from the front seat. The two of us in the backseat are along for the ride to offer our highway expertise, gleaned from past culturally sensitive road projects like U.S. 93 in Montana and Paris Pike in Kentucky. Together, our task is to propose solutions to the state's highways in the face of climate change and the inevitable generic-nature of roads – to make them, in a sense, more Hawaiian. We stop every couple of miles at key points – the Banyan tree, a state park with a fishpond, a narrow stretch of road below the cliffs with the profile of the Crouching Lion above us. At each stop we take notes of scenery, places of cultural significance, circulation patterns and the health of the vegetation.

As we pile in the car, the local architect hesitates at the open door and asks me "What did you see?" Not knowing how to answer, I ask him what he means. "What exactly are you looking for when you're clambering around the embankments and taking notes?" I pause, overwhelmed by the thoughts flickering through my mind …

I think of the maps we had compiled showing the boundaries of the ancient *ahupua'a* that structured traditional Hawaiian culture. A series of *ahupua'a* stretch along the windward coast here bounded by mountain ridges. The Hawaiian meaning is similar to "watershed" but connotes

0.2 Collage of *ahupua'a* along leeward Oahu.

an abundant source of livelihood to its inhabitants. Each *ahupua'a* had a chief who distributed land, produce and fish to its peoples in a rigid hierarchy of social and economic class. An island's *ahupua'a* combined to form a *mokupuni*, the largest designated landform (*moku*) surrounded by water (*puni*). Up until Kamehameha in the early 1800s, each island had shifting kings moving in and out of power, so it was the *ahupua'a* and its chiefs that remained constant. A reliable source for all that was needed in life.

I recall the previous day's workshop led by a native Hawaiian elder who worked with us to envision a more inclusive Oahu. She often worked in conflict resolution between those *haoles* (people of European descent) who bring ideas to improve landscape, infrastructure or social systems and native Hawaiian peoples who understandably resent the imposition of other values and ideas. She was kindness itself but gently chastised our "expert" knowledge for its assumption of comprehensiveness.

I remember the tour of *loko i'a* (fishponds) led by a local historian who explained the process of Hawaiian aquaculture and the preservation efforts underway. The *loko kuapa* (low stone wall) could still be seen running out into the bay off the windward coast.

I, along with my companions, have already made five stops on this stretch of road, the red earth staining the edge of the asphalt roadway near the Valley of the Temples, the interstate's flying superstructure inflicted on some of the most scenic landscape in the world, the gas station with three palm trees framing a view up the coast.

I think of what I do not see because I am not of this place. In a colonized island like Hawaii, the important things are often hidden or missing. The lost art of feathering a royal dress. The lost birds (now extinct) used in the feathering. The native peoples who would reside in this *ahupua'a* if not for two centuries of white man's disease and decadence. The voices of those lost people who might speak about this place and the effects of the highway. The pollution from the road, deposited by the cars of the tourist and native alike, slowly migrating across the highway along the roadside and into the groundwater. The senses other than sight that speak to us through the wind in the palm trees or the smell of a gardenia.

I think of the swim in the ocean I had taken the night before … a bewildered attempt to swim as far out as I could, only to prove to myself I was no great swimmer or island dweller. And I responded to his question with a wave of my hand meant to encompass the sweep of the landscape and my own inadequacies but unfortunately with the appearance of dismissing his query.

* * *

"What do you see?" The question embraces more than sight, but the close relationship between observer and surroundings. It is an important question for designers or anyone who shapes the landscape. The question suggested I saw something different than the others, but of course, I was a visitor from the mainland with my own set of conceptions/images of Hawaii I had brought with me. Yes, my expertise in highway design and cultural landscapes offered a different and useful point of view. But I struggled to articulate what I saw, all the experiences even of the short five days we had spent in Oahu, to frame the process of site analysis, to offer steps to an understanding of the highway's influence on the landscape and the landscape's influence on the highway.

The question followed me when I returned to Seattle, as I traveled throughout the western United States analyzing places, meeting with communities and designing landscapes. I continued to ask it as I did

0.3 Highway 93, Waianae, Hawaii.

fieldwork for a PhD in geography on urban public space and homelessness. And I continue to ask it of my students in landscape architecture and sustainable environmental design: What do you see? What stands out to you about this place and how will you record it, represent it? This book is my answer: to set down some ideas and comprehensively evaluate how designers examine the landscape, offer some critique of existing processes of analysis and potential ways to generate a foundation for more inclusive, creative and sympathetic design.

We drive another mile along the highway, the driver pulls over off the road to the side of what looks to be a shack made of corrugated tin. Getting out, we walk into the heat along the narrow shoulder of the road to line up in front of an opening in the shack next to the Waihole Poi Factory. We order cups of *poi* and some *haupia* (coconut pudding). Scooping out the mild paste of the taro plant in the shade, leaning on the car by the side of the road, the architect states "This is Hawaii." I agree; I am the *kōlea*, gorging myself on the scenery, the hospitality and the poi, only to return fat and rejuvenated to the mainland … with a different point of view.

SITE ANALYSIS

When we analyze the landscape, what is it we are doing? We perceive our surroundings through multiple senses: seeing, hearing and touching the environment. We form impressions of a place filtered through past experiences of other places both similar and different. We focus on certain landscape elements based on an aesthetic response, both positive and negative. We observe how others interact with a space. While some of our perceptions stem from knowledge of social and ecological systems, these do not necessarily form the core of our analysis or our understanding. We take impressions of the landscape and our scientific and not-so-scientific understanding of a place and draw them, write them down as notes, take them back to an office or home, and let them stew. We may engage in a systematic process of understanding, dividing the landscape into layers and relating those layers to other layers. We may do this digitally or by hand. Ultimately, we compile information into a base or foundation we will use to guide design.

In the professional world of landscape design, this process happens in the context of economics and efficiency. Clients ask professionals to design a space with a certain budget and scope of work. If a landscape architect has six hours "on site" to reflect on landscape, observe its use and draw up preliminary ideas to use as a base for design, then efficient metrics and analytics must be used. It is possible to learn about a place without ever going there. Travel is expensive and energy-intensive. If site work can be minimized at the beginning, project costs can be lowered. Designers have turned to remote information databases, such as Google Earth and Google Street View, to build a contextual picture of the landscape and replace some on-the-ground initial work. Valuable tools for processing information and showing relationships (primarily from above) sort information quickly, yielding clear interpretations of what can be incredibly complex data.

For the Hawaii Roads project, our office in Seattle was an ocean away from the highways of Oahu. So before flying there to drive the roads, we used a Geographic Information System (GIS) to map *ahupua'a* boundaries, the sites of culturally significant landforms (according to the State), patterns of vegetation and patterns of development. We assembled a framework of analysis, so that when we drove the highway we could locate ourselves in the landscape according to both traditional Hawaiian landscape patterns and contemporary development.

For all of the benefits of GIS and other mapping tools, a top-down representation of a landscape is not a landscape. As Alfred Korzybski, a Polish American philosopher, stated "the map is not the territory."[1] Strategic planning from a bird's-eye view suffers from an inability to see detail, a blindness to the movement of people on the ground, and a privileging of the visual, which can result in perpetuating existing forms of inequity.

Representation, itself, has been critiqued as too divorced from reality to form an adequate basis for research and decision-making.[2] We will not rid ourselves of representation, as it is part of dialogue, but representation (from above) rests on a number of assumptions:

1. that a representation of a place accurately represents that place (accuracy);
2. that a top-down view is the best way to organize information (relational);
3. that the representation captures the various elements necessary to make future site decisions (adequacy);
4. that a visual understanding of a place is enough (sensory); and
5. that the landscape will remain static like a map and not change, i.e. weather, growth, flows, development (dynamism).

A map should never replace immersion in a place.

Site analysis – the disassembling of a place into separate elements, evaluation of said elements and reassembly to draw conclusions about the landscape – mirrors a uniquely human way of comprehending the world by breaking it up into parts that can be better understood. A planner, in an attempt to learn about a place, may inventory the water on site: channels, water storage, amount of runoff at various times and potentially the groundwater below the surface. She would analyze the quantities of flow to evaluate the health of the hydraulic system and thus inform a new landscape design. Or, more likely, she would rely on a hydrologist's report to summarize these processes, a disassembly of disciplines in parallel to the division of the landscape into layers.

According to James LaGro, site analysis assesses a site's suitability to meet human needs in a sustainable way.[3] It provides information to make good decisions about land use. Although not always articulated as a process of division, site analysis can be broken down into the following steps:

1. Site selection and programming.
2. Separation of the landscape into parts (i.e. cultural or natural).
3. Inventory of these different parts.
4. An analysis of these different parts – are they healthy? Working?
5. Synthesis of these parts into a basis for design.

These steps can work iteratively, as data is gathered, shaping and informing both prior and anterior steps, although often it is done linearly.

The site analysis method has various advantages. It divides a landscape into various systems or elements which have their own extensively developed areas of expertise. The landscape elements can be divided in such a way to facilitate measurement, offering a quantifiable way of determining ecological health. For instance, hydrology offers methods for analyzing the quantity and quality of stormwater flows. Water quality can be assessed for the amount of pollutants (usually in parts per million (ppm)). Discharge from a channel can be quantified over time using stream gauge instruments. Landscape divisions lend themselves to mapping through a process of layering each division on top of each other, working from above. To draw meaningful conclusions, the reassembly of the layers must convert divisions into some consistent measure of suitability or health or susceptibility to meaningfully compare different areas, i.e. high, medium and low suitability. If consistency is achieved and values are assigned, sensitive impacts can be avoided.

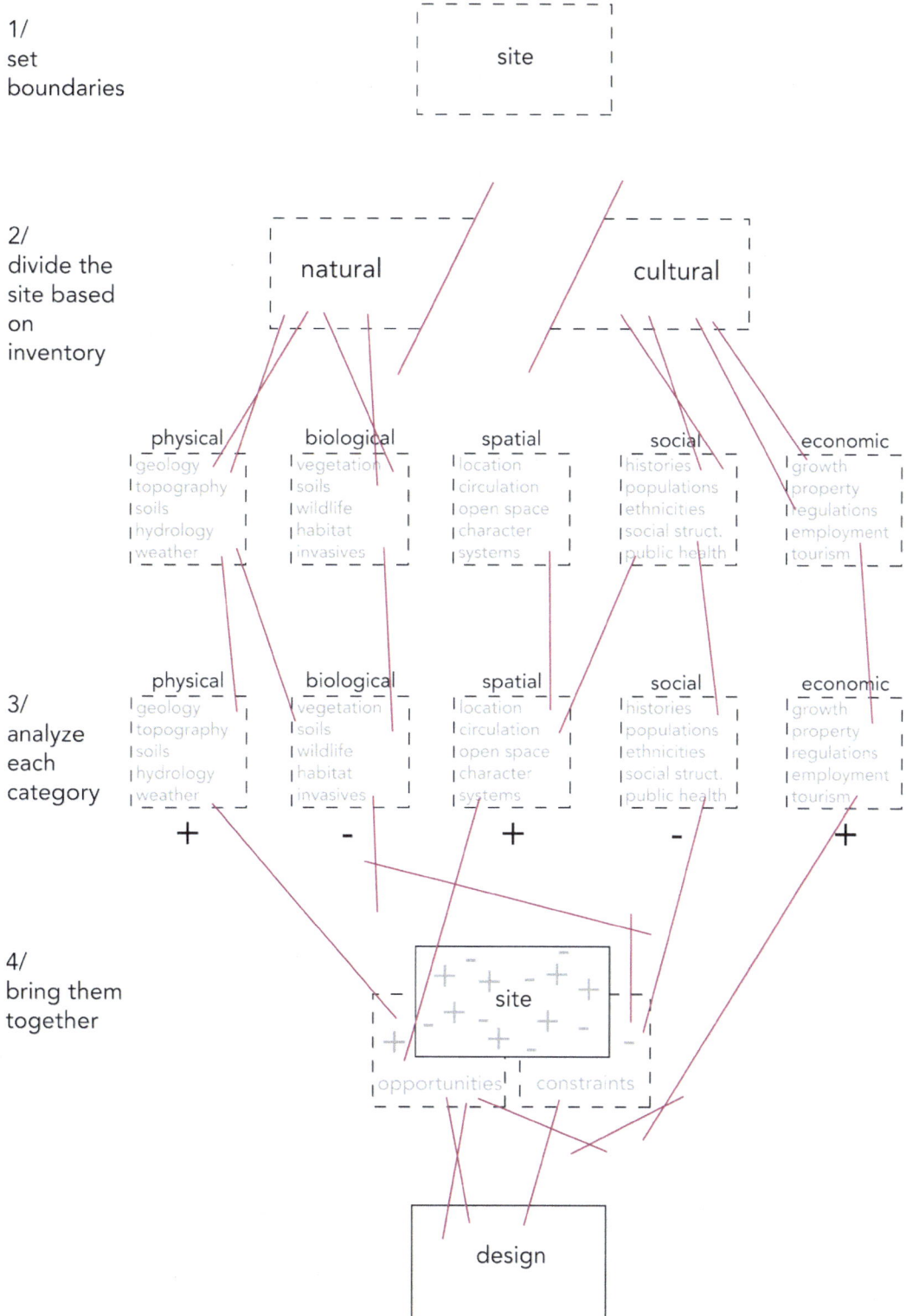

0.4 A diagram of the site analysis process.

Yet by chopping the landscape into parts, site analysis can miss the whole. The whole is a spatial thing, a systems thing. Spatially, the designer limits landscape to site, a bounded portion of the earth, a social construct of partitions. The word "site" conveys both a static given and a contained space. Neither is an accurate referent for landscape. Even when bounded with a fence or wall, landscape has no respect for our lines of demarcation. Water flows across and through. People hop a fence. Gophers burrow a warren of dens. Wind blows. Earth erodes. Landscape is neither static nor contained. The term "site" prioritizes property, the rule of law, and limits our responsibility as designers to a little world.

The other half of site analysis also narrows the process of landscape evaluation in an effort to manage a complex interaction. While the landscape flows, morphs, and reconfigures itself temporally, analysis disassembles these flows to understand and simplify intertwined, socio-ecological systems. Just as the landscape cannot be bounded in a meaningful way, analysis struggles to address cross-boundary, systems thinking. Water, plants, soil, people – elements integral to the vocabulary of landscape – move within the landscape and frustrate efforts to clearly delineate where one ends and the other begins.

The practice of conventional site analysis has long-term consequences for our landscapes:

- A reliance on measurements misses the difficult-to-quantify, amorphous aspects of a place.
- An assemblage of layers of information neglects the spatial and form-making properties of a place, particularly if compiled in plan.
- A reliance on suitability as the driver of synthesis can create "least-bad" alternatives and miss the potential for creativity.
- Gathering information from a site separates the designer from the landscape, leading to assumptions of (false) objectivity.
- Bounded places miss context, the incredible flows and networks of moving and breathing earth, creating places and buildings without any relation to their surroundings.
- A focus on property boundaries perpetuates existing systems of inequity and exclusion, cutting off the voices of those without access to property.

Each of these limited ways of thinking and resultant consequences will be addressed in later chapters. For now, a few examples will suffice.

In the 1980s, George Hargreaves, then with the SWA Group, designed an office plaza between two modernist office buildings in Denver, Colorado called Harlequin Plaza. The design consisted of a skewed grid of black and white tiles laid out on the surface of the plaza, backed by pink and mirrored walls meant to transcend site by invoking the shape of the Rocky Mountains. Pictures of the landscape received awards in *Landscape Architecture* magazine. It was described as a bold embrace of modern/post-modern/Cubist design (depending on the critic). Yet ten years after installation, office workers were tripping on peeling tiles, tile colors had faded to dark and light grays, and visitors complained about the glare from the sun as they approached the building. The contractor blamed the designer; designer blamed contractor. And eventually another firm redesigned the site using more appropriate materials.

The cultural elites at the time collectively lost their minds over Harlequin Plaza, calling it "a consequential work signaling new directions"[4] and lauding it for its mountain and midwestern grid referents,[5] apparently not noticing it was ugly, dysfunctional and unsustainable, a trifecta of failure. Branding aside, Denver sits on the edge of the Great Plains, not in the Rocky Mountains. The mountains' distance weakens the potential

0.5 Harlequin Plaza in Denver, Colorado in the 1980s.

metaphor of the pink slabs of metal. Designers ignored the needs of the office worker, instead confronting them with a skewed grid with an unclear purpose. Designers did not consider weather as an integrated element of the landscape. Black tiles absorb heat; white tiles reflect it. The freeze and thaw of the high plains, the close, high altitude sun require thick durable materials. A better approach to landscape engagement might have embraced the site's specificity and context. Although this may not have shifted the designer's focus from magazine cover glory to the people/landscape, it may have inspired an internal debate before project design rather than the external debate focused on controversy and prominence in which people and landscape were sidelined.

Contrast this with another mountain landscape – Icicle Creek Conference and Music Center in Leavenworth, Washington – a project designed by John Paul Jones of Jones & Jones Architects and Landscape Architects. The design nestles a series of conference buildings and landscape "rooms" within a mountain setting. A traditional site analysis of the wooded location was supplemented with visioning sessions on what it means to "retreat," to get away from the everyday and think clearly. The resulting design emphasized the building/mountain metaphor, but in a more direct experience that could be understood by visitors. Local materials like Douglas fir siding integrated the project within the forest. Steep metal roofs adapted to the high snow loads of the Cascade Mountains crown each of the buildings. Visitors to the conference center have views and facilitated, interconnected spaces to rejuvenate their career/business/mind.

0.6 Icicle Creek Conference and Music Center, Leavenworth, Washington by Jones & Jones.

In a different milieu, the High Line project designed by James Corner and Field Operations has been discussed extensively in the design literature.[6] The City of New York and Friends of the High Line had the inspired idea to convert an elevated railway into a linear park. Corner did an excellent job incorporating the old rail infrastructure into the elevated path. Native plants chosen by Piet Oudolf reference the ecological history of New York. The site analysis must have been extensive. However, the project's success and popularity raised property values along the elevated corridor, driving low- and middle-income people from the neighborhood. For many, the High Line is now a symbol of gentrification. Would the design be different if they completed a socio-economic analysis of the project's effects? Would the project have proceeded at all? It may have been impossible to predict the consequences, but it would seem that gentrification may have been the goal. Field Operations marketed the project as a "powerful catalyst for investment."

In design we activate a space for certain people, while others are left out – teens, older people who no longer drive, people experiencing homelessness. Asking the question "who is this place for?" situates the design of space for somebody within a political context. "For somebody" most frequently refers to the production of space for the wealthy, for economic gain or for pleasure of the few. The practice of site analysis and behavior mapping learned in school, a practice of recording human activity on the site in hopes of redesigning that site to increase users of the space, can lead to more of the same, an increase in the inequity of space, if the composition of spatial users is not addressed. At best, it leaves us well-regarded, human-centered designs to be enjoyed by the wealthy, i.e. Paley Park in New York City.

Weather, office workers, materials, social justice – site analysis must incorporate so many different factors before reassembling them in a meaningful way, it is no surprise it often misses the mark. Scholars from other disciplines have critiqued the site analysis process.[7] If site analysis fits within the broader skill set of reading the landscape, geographers have wrestled with the implications of this reading, in particular *who* is reading and *from where* they are reading. Cosgrove et al., along with others in the cultural landscape tradition, suggest the landscape is like a text, full of symbols reflecting societal values.[8] The ability to read this text enhances our ability to navigate the landscape. This is an intriguing idea, gesturing towards a landscape-reading-as-skill that could be developed by designers. Gillian Rose qualifies the idea of reading the landscape, arguing the reader's point of view is almost always male, from far away and thus couched in "objective" terms.[9] Thrift argues for a non-representational approach to landscape, acknowledging the inability of us to understand it from afar and to represent it in all its wholeness and complexity.[10] Each theoretical position contextualizes the act of reading or listening to the landscape.

Whether from a position of privilege or not, reading the landscape is central to what planners, architects and landscape architects do, at least in the sense of trying to understand its situation and processes. In the face of challenges to site analysis and extracting information, a better metaphor of understanding the landscape is conversation. We either listen to the landscape or talk over it. All too frequently, we inhibit conversation with the land. We discuss instead of dialogue.[11] Discussion splits us into camps or corners from which opinions are defended and truths are argued. When we defend our own opinions, we stop listening and refuse to understand others' opinions and ideas. Clues to solving an entrenchment in a positional analysis of landscape lie in this examination of the use and process of language.

DIALOGUE

In his book *Standing by Words*, Wendell Berry links the disintegration of community with a disintegration of language.[12] Berry connects language to our relation to landscape, in his case, the agricultural landscape of western Kentucky. In 1964, he quit his position as a professor of English literature at New York University and moved back to Kentucky. In 1965, he and his wife purchased a farm called Lane's Landing near his childhood home. As he argues, care for the farm led to a care for words. In contrast, a lack of care of the land has rendered contemporary language incomplete and, as a descriptor of things, incomprehensible. It no longer designates its object precisely. Its speaker does not stand by it, believe it or act on it.

According to Berry, poor language takes two different forms: sentimental generalities found in romantic poetry and objective "expert" speech found in environmental impact reports. In contrast to both, good speech is both precise and personal. It is the speech of people living in the same community. Words point to something specific, visible or remembered. "The hollow beech tree blew down last night."[13] If speech needs to be abstract, it is the rigorous speech of tension, of conflicting ideas. It wrestles with the complexity and paradoxes of life.

Following Berry, landscape dialogue is precise and personal. It listens to the person/landscape to hear of the specific concerns/challenges/triumphs. It speaks with precision, designating particular places and processes as ongoing rhythms and entities experienced by specific people.

To transcend site analysis, as it exists in the disciplines of architecture, landscape architecture and planning, requires dialogue. The speaking of words is the continuous response to relations and signifiers reflecting the complexity of real life. In conversation with others, we understand received ideas incompletely. Spoken words and phrases mean different things to different people, particularly if the concepts are abstract. Psychologists describe one simplification of conversations as "listening with intent to reply."[14] We hear what others have to say only in order to form a reply in our minds, then wait for our turn to speak. The result: a lack of understanding of the other. In a similar way, designers visit a site only long enough to form an idea of a design (in this metaphor: the speech act). Understanding of landscape remains superficial. A mere glance (or photograph) informs design. Dearly held notions of culture or nature influence the design, rather than an openness to a specific landscape's possibilities. Understanding may be hindered by a reliance on digital "access" to a site in which a glance is all that is available.

Unlike the speech of the expert or the sentimentality of the poet, a conversation with the landscape occurs within a concrete, human-inhabited place, not amongst abstract ideas. Our understanding will always be incomplete, garbled or cursory. However, we can engage with

0.7 Dialogue requires immersion in the landscape, New Orleans, Louisiana.

each place's specificity, each place's personality. We can ask the following questions:

- How can we engage with the substance of place rather than its trappings?
- How do we interact with landscape in a meaningful way, as part of ourselves, rather than as a separate object?
- And what kind of impact will a deeper, more specific, engagement with that place provide?

Landscape dialogue is the practice of listening to the space around us. It is the everyday engagement with the world, an awareness of the connectedness of things. Whether on an urban sidewalk or a rural road, landscape dialogue ensures an honest and open acknowledgement of the close relationship between ourselves and the landscape.

A landscape dialogue approach uses the metaphor of language, not to read the landscape, but to be transformed by it. The landscape speaks for itself. It responds to our trajectories; adapts to our "restorations," coalesces into place, into a whole to be experienced as a whole. "Site," while a useful concept at times, does not exist in the landscape. Context is not something to be added to a project; it is the project. Landscape is the relation between things, between ideas. It exists as connections, networks, a web of interactions. We try and read by first listening (using the other senses); we try and taste the landscape, and only then do we see. We speak to test out ideas, to voice our concerns, to give voice to the inequities we have embedded in the landscape. Ultimately, then we read the landscape – not as a gatherer of information in order to control but as a dialogue to be transformed.

Designers have a choice, just as we do in meeting and interacting with others, we can argue and defend dearly held opinions or we can listen, try and understand and participate in the co-creation of the landscape. The listening approach connects with others to find new solutions as fellow participants, as David Bohm states:

> In dialogue, when we have a very high energy of coherence, it might bring us beyond just being a group that could solve social problems. Possibly it could make a new change in the individual and a change in the relation to the cosmic ... It is a kind of participation. The early Christians had a Greek word koinonia, the root of which means "to participate" – the idea of partaking of the whole and taking part in it; not merely the whole group, but the whole.[15]

Table 0.1 Comparison of a landscape dialogue approach to site analysis.

Conventional site analysis	Landscape dialogue
A linear process	An iterative, in-depth process
Seeks information to understand the landscape	Seeks transformation through the landscape
Landscape as object	Landscape as subject
Analytical approach	Critical and immersive approach
Problem-solving mentality	Openness to transformation

Landscape dialogue, resting on the foundation of a relational view of the world, is partaking of the whole. It is the process of making the engagement with landscape visible, making it known, so that we can address the relationship. This direct engagement with the landscape does not preclude the idea of landscape as social construction, it just removes the option of describing landscape as something "out there" that we can view from above, as we do "culture" or "environment." All descriptions of landscape come from a place of inhabitance within that landscape, making it impossible to be objective in the scientific sense of being an outside, impartial observer.

Given the entanglement of landscape and ourselves, the analysis of a place/site/landscape is recognized as a qualitative conversation. Quantitative measurements will be made, but these ultimately support a qualitative assessment of both the landscape and our relationship to it.

Other have explored the idea of landscape as dialogue. A whole body of work equates architectural elements with words.[16] In this book, however, I want to examine dialogue with and of the landscape as a dynamic process … not a static assignation of symbol to object.

The most extensive effort of landscape-as-text is Anne Spirn's *The Language of Landscape* in which she argues for an understanding of the landscape through a better reading of place. She takes a tour through significant landscape projects, discussing environmental tensions, many the result of misreading a landscape's language. Spirn suggests "landscape is the scene of life, cultivated construction, carrier of meaning. It is language."[17] I draw from her literate engagement with the landscape. Yet the process of landscape evaluation is less a reading of words and more a listening to its rhythms and a speaking to its potential. Understanding the flow of a creek or the movement of geese across the sky will not yield, in and of itself, more sensitive designers, more ethical civic leadership and a more engaged community. Designers have so much information available to them compared to the past and it has not necessarily led to more sensitive ecological planning or social justice. We engage the landscape from within it, not as experts or seekers of information, but as inhabitants. Spirn insists landscape dialogue is an action, or more appropriately an interaction – the building of a dike, the planting of hedgerows, the painting of a garden – and an engagement with others – through public meetings, through a conversation with a client, through a class visiting a landscape. It is this dialogue, rather than landscape as language, that opens us up to transformation. It is a subtle but important distinction.

Stilgoe's *What is Landscape?* offers an even deeper dive into the words of landscape. His lexical research relies on descriptive words of the farmers, sailors and laborers of northern Europe, whose words did not make it into the *Oxford English Dictionary*, but nevertheless evoke the historical change over time of landscape. He links the word "landscape" to a child moving sand with a pail and shovel. From the Netherlands, it is the earth "thrown up against the sea." Stilgoe gives a thoughtful description of landscape dialogue:

> Each observer creates a concatenation of space and structure peculiar to himself or herself simply by noticing. Frustration originates in analysis, in questioning, in hearing explanation accurately, in words and in the silences in which inquirers and the inquired-of grapple for words, for the right words.[18]

Analysis is useful but distancing, like receiving advice instead of empathy. There is a time and place for analysis. It needs to be done in such a way that maintains connections, maintains relations between landscape and between people.

The workings and practice of spatial engagement with place shape landscape dialogue, *the purpose of which is to transform not only the landscape but the designer as well.* As designers, we engage with a place to not only understand it, but to respond to its callings to become more sensitive, more alert, more inspirational people of change. To change the landscape for the better, to regenerate healthy landscapes, requires a processing of information about the landscape, how it works and how our immersion in it influences landscape health. But landscape dialogue moves beyond an informational approach to listen, understand and transform a place. It is ultimately less about gathering information, driving back to the office and breaking the landscape into its constituent parts, to reassemble them later; it is more about listening to a landscape, speaking into that landscape, to transform both it and ourselves.

This book builds on over 20 years of landscape architecture experience of dialogue to relate methods, articulations, exercises of understanding the landscape, and in turn using that information to inform design and transform ourselves. The book does not outline a linear method of inventory, analysis, program and design found in offices and studios. As the designer engages in inventory and analysis, it is important to ask more fundamental questions:

- What is our role in the landscape?
- How do we evaluate the landscape and use that evaluation to make design decisions?
- How can we do this in a positive manner for transformational power and landscape health?

As we answer these questions, we find a need for a more immersive process of landscape design.

As a designer of landscapes, I frequently use the royal "we" to indicate the book's audience of architects, landscape architects, planners and civil engineers who actively shape the landscape through design and policy. Each of these disciplines has its own process of landscape analysis (however cursory) and each would be enhanced by a more direct engagement with place. The ideas of landscape dialogue will be received differently by different professions. Those disciplines, such as engineers and planners, more embedded in existing systems of power dynamics, in the urban growth machine,[19] may discover a need to see context, to seek out connections between policy and urban form, to see the landscape as a product of a specific history. Other disciplines, more peripheral to the political process, may be inspired to spend more time engaging with places to transform themselves. My experience as a landscape architect, first in public practice then at a private architecture firm, suggests the need for a more holistic way of looking at landscape, inspired by design, but also by non-design disciplines of anthropology, sociology and geography.

The purpose of the book is three-fold:

1. to challenge (without necessarily replacing) conventional site analysis;
2. propose a dialogue with the landscape that emphasizes listening, immersion and self-transformation; and
3. to examine ways of interacting with the landscape as a part of ourselves.

THE STRUCTURE OF THE BOOK

Each chapter begins with an idea/theory. I draw from inter-disciplinary thinkers to wrestle with ideas on landscape. The bulk of each chapter is then a discussion of the experience of landscape – the at times distant, at times intimate, interactions between designer and the landscape, and the meanings associated with this experience. In the final section of each chapter, I discuss a few methods of putting these ideas into practice – an application of theory or praxis. How does a landscape architect or planner experience the landscape in focused and practical ways so that the designer/planner can make decisions on healing and accommodating people? And how can these qualities and experiences be represented to others?

A deeper engagement with landscape must address the following concerns:

- *Landscape dynamism* (Chapter 1). The problem of "site." I first approach the *subject* of analysis. What is it the designer evaluates? A visual scene? A static placement of elements? A stage where human drama plays out? A site? I suggest that the primary quality of landscape is its changing nature – the shifting relationships and flows that weave together a milieu of space. If landscape is dynamic, design must be dynamic.
- *Landscape information* (Chapter 2). The problem of "analysis." I then address the *process* of analysis, conventionally the accumulation of data and measurement in the pursuit of objectivity. What is data on the landscape for? How do we process this steady stream of information? What are the limits of analysis? This chapter argues for a dialogic approach to landscape analysis instead of data gathering. Information should not be ignored but challenged as part of the landscape itself.
- *Landscape perception* (Chapter 3). If conventional ways of assessing the landscape fall short, how then do we see, feel, embrace the landscape as the landscape? In this chapter, I engage with current research on perception to challenge the idea that humans perceive a static picture. We perceive by moving in and around the landscape, resulting in a rich and textured experience of place, not as a separate entity, but as a connected participant.
- *Landscape immersion* (Chapter 4). Absorbing a place's gestalt. If perception is an immersive process of movement, what does it mean for designers to move in and through a space? How can we translate immersion to form a basis for design? In contrast to data gathering, we experience the landscape within the landscape and then communicate

these experiences to others. The key element in this process is time in the landscape.

- *Landscape relations* (Chapter 5). The space between us. How do we experience the landscape, not as isolated individuals but as people within a socio-ecological space? How do we account for others in landscape dialogue, particularly those different from us? Connections between people and spatial elements structure a place. Designers maintain an ongoing interaction with a network of people and spaces in order to evaluate a space.

- *Landscape critique* (Chapter 6). Landscape as a network of shifting relations acknowledges that these relations might be uneven, that is, certain people or structures impose "standards" and rules on others in discriminatory and even violent ways. How should designers account for historic imbalances in the landscape? How do we assess power imbalances as intrinsic within the contemporary landscape? I argue for a critical approach to landscape evaluation so that alternative stories and voices enter the dialogue. Critique requires courage to confront existing practices and our own culpability in inequities.

- *Landscape dissent* (Chapter 7). Transgression and protest as dialogue. Critique of power in the landscape is not enough. How do we as designers understand and influence shifting power in the landscape in an experiential way? As an act of dialogue? This chapter extends the last chapter's focus on critical landscape evaluation to include immersive techniques of protest and dissent as a dialogic tool of landscape evaluation.

- *Landscape formation* (Chapter 8). The landscape reads us. Dialogue is a conversation, both listening and speaking. In that dual exchange lies the potential for personal transformation. If we, as designers, do not change along with the landscape, as part of the landscape, our design efforts become stilted and static. In the culmination of the argument for landscape dialogue, this chapter urges an openness to personal formation into more inclusive and creative designers.

On the Kamehameha Highway along the windward coast of Oahu, the designer can drive through each *ahupua'a* linking mountains to ocean, to record in a myriad of ways the experience of movement in the landscape. Immersive perception fails, by itself, to portray the complete experience of the place. The visitor must also know how highways work, the engineers' reliance on applying American Association of State Highway and

gradient of natural resources

0.8 Traditional resource management from mountain slopes to the ocean – *ahupua'a* on Oahu's leeward (dry) shore.

Transportation Engineers standards developed for distant places and far away roads. The relationship between the highway's landscape and future climate change and rising oceans. The displacement of native peoples from the lush valleys on the windward side. The potential for dissent through marches, jaywalking, eroding banks and stalled cars on the side of the road, all speaking from under the island's choking necklace of pavement. The landscape crying out. It speaks to us.

PRAXIS

Like other texts on landscape and design, I encourage designers to keep a sketchbook to assess the landscape visually. There is no better way to assess the spatial qualities of a place than the habit of regularly drawing in the field, whether that field is a threatened watershed or an urban street. A few points on the practice of sketching:

1. Develop a practice of sketching in everyday places. Do not wait until you sit before the Eiffel Tower. The more urbane and ubiquitous, the better. This may mean that designers carry a smaller sketchbook for everyday use to sketch in the many pauses in the day.
2. While some designers publish beautiful sketches of landscapes, the primary purpose of sketching is to understand space through the embodied practice of drawing on paper (or iPad). Do not be concerned with the quality of the drawings. Practice patience with one's technique.
3. Sketching is the designer's note-taking. As such, fill the page with drawings and notes. Annotate everything, particularly if it is quicker to ask questions of a space rather than draw potential spaces.
4. Go back to prior sketchbooks periodically. I am always surprised at prior insights on contemporary issues found in my sketchbooks that I had forgotten. Build a visual and spatial understanding of everyday landscapes.

For guidance on sketching, consider Chip Sullivan's *Drawing the Landscape* or Janet Swailes' *Field Sketching and the Experience of Landscape*. Many of the exercises I suggest have a sketching component in an attempt to record the experience of landscape while one is in the field. Landscape perception, immersion and critique all rely on sketching to understand the landscape and communicate that understanding to others. Sketching becomes the speaking component of landscape dialogue to be encouraged, developed and presented. Sketch everywhere.

Teaching notebooks, 5 x 7" for design and landscape architecture, one per class

Sketchbooks, 5 x 7" spiral bound for sketching in the field the spatial characteristics of the landscape

Solid gold owl

Small notebooks for field notes during observations of people experiencing homelessness in the landscape (thesis work)

Notes and sketching, 5 x 7" notebooks for ideas, diagramming relationships and recording theoretical positions; Notes on readings

0.9 Author's "stealth" field notebooks from dissertation research, larger design sketchbooks and teaching notebooks with written and visual ideas.

NOTES

1. Alfred Korzybski, *Science and Sanity: An Introduction to Non-Aristotelian Systems and General Semantics* (New York, NY: International Non-Aristotelian Library Publishing Company, 1933).

2. Ben Anderson, *Taking-Place: Non-Representational Theories and Geography*, ed. Paul Harrison, New edition (Farnham: Ashgate, 2010).

3. James A. LaGro, *Site Analysis: Informing Context-Sensitive and Sustainable Site Planning and Design*, 3rd edition (Hoboken, NJ: Wiley, 2013), http://site.ebrary.com/lib/ucdavis/Doc?id=10653568.

4. Marc Treib, "Complexity and Spectacle: Three American Landscapes of the 1980s," *Journal of Landscape Architecture* 10, no. 3 (September 2, 2015): 6–19, https://doi.org/10.1080/18626033.2015.1094899.

5. Anne Whiston Spirn, *Language of Landscape* (New Haven, CT: Yale University Press, 2000).

6. For example, Kevin Loughran, "Imbricated Spaces: The High Line, Urban Parks, and the Cultural Meaning of City and Nature," *Sociological Theory* 34, no. 4 (December 1, 2016): 311–334, https://doi.org/10.1177/0735275116679192.

7. Kenny Cupers, "Towards a Nomadic Geography: Rethinking Space and Identity for the Potentials of Progressive Politics in the Contemporary City," *International Journal of Urban and Regional Research* 29, no. 4 (December 1, 2005): 729–739, https://doi.org/10.1111/j.1468-2427.2005.00619.x; Don Mitchell, "New Axioms for Reading the Landscape: Paying Attention to Political Economy and Social Justice," in *Political Economies of Landscape Change: Places of Integrative Power*, ed. James L. Wescoat and Douglas M. Johnston (Dordrecht: Springer Netherlands, 2008), 29–50, https://doi.org/10.1007/978-1-4020-5849-3_2.

8. Denis Cosgrove, Stephen Daniels and Alan R.H. Baker, *The Iconography of Landscape: Essays on the Symbolic Representation, Design and Use of Past Environments* (Cambridge: Cambridge University Press, 1988).

9. Gillian Rose, *Feminism & Geography: The Limits of Geographical Knowledge* (Minneapolis, MN: University of Minnesota Press, 1993).

10. Nigel Thrift, *Spatial Formations* (London and Thousand Oaks, CA: SAGE, 1996).

11. David Bohm, *On Dialogue* (London and New York, NY: Routledge, 1996).

12. Wendell Berry, *Standing by Words: Essays* (San Francisco, CA: North Point Press, 1983).

13. Berry, *Standing by Words*, 33.

14. Marshall B. Rosenberg, *Nonviolent Communication: A Language of Life* (Encinitas, CA: Puddledancer Press, 2003).

15. Bohm, *On Dialogue*, 41.

16. Umberto Eco, "2. Function and Sign: Semiotics of Architecture," in *2. Function and Sign: Semiotics of Architecture* (New York, NY: Columbia University Press, 2019), 55–86, https://doi.org/10.7312/gott93206-004; Donald Preziosi, *Architecture, Language and Meaning: The Origins of the Built World and Its Semiotic Organization* (Walter de Gruyter, 1979).

17. Spirn, *Language of Landscape*, 15.

18. John R. Stilgoe, *What Is Landscape?*, Reprint edition (Cambridge, MA: The MIT Press, 2018), 51.

19. Gordon MacLeod, "Urban Politics Reconsidered: Growth Machine to Post-Democratic City?" *Urban Studies* 48, no. 12 (September 1, 2011): 2629–2660, https://doi.org/10.1177/0042098011415715.

The landscape speaks: the problem of "site"

Landscape is a place of overlapping flows, not a static container of space.

On a cool Friday afternoon in March of 2011, the Honshu plate, responding to 1,100 years of accumulated stress from pushing against the larger Pacific plate, abruptly slipped, lifting the floor of the Pacific Ocean upwards by several meters. The plates' displacement shifted the islands of Japan three meters closer to North America. Primary shockwaves called P waves, moving through rock at five miles per second, radiated out from the epicenter through nearby Japan. The short wavelengths did little damage. Secondary shockwaves called S waves followed the P waves. With longer, more sustained wavelengths, S waves destroy buildings. They reached Sendai, Japan at 2:46 pm and Tokyo at 2:47.[1]

The resulting earthquake lasted for six minutes.

Once the trembling stopped, there was a pause. People on the island nation of Japan understood the potential for a tsunami to follow. Japan has the most sophisticated earthquake and tsunami warning system in the world. Within nine minutes of the earthquake, alerts flooded cell phones with a tsunami warning for all of Japan, an estimated three-meter-high wave would hit the east coast 30 to 45 minutes after the earthquake. Japan's Meteorological Agency based the predicted wave height on an initial earthquake measurement of 7.9 magnitude; in actuality the earthquake was 12 times stronger.

To prepare for just this catastrophe, the Japanese had modeled the impact of a possible tsunami to determine who would be at risk and who would not. Based on these models, the city had constructed embankments along the coast, built a six-meter roadway berm along the Sendai coastal plain and established evacuation hills, temporary centers and shelters throughout coastal neighborhoods. When the tsunami arrived on that March afternoon, these defenses failed. They did more than fail; they

DOI: 10.4324/9781003158943-2

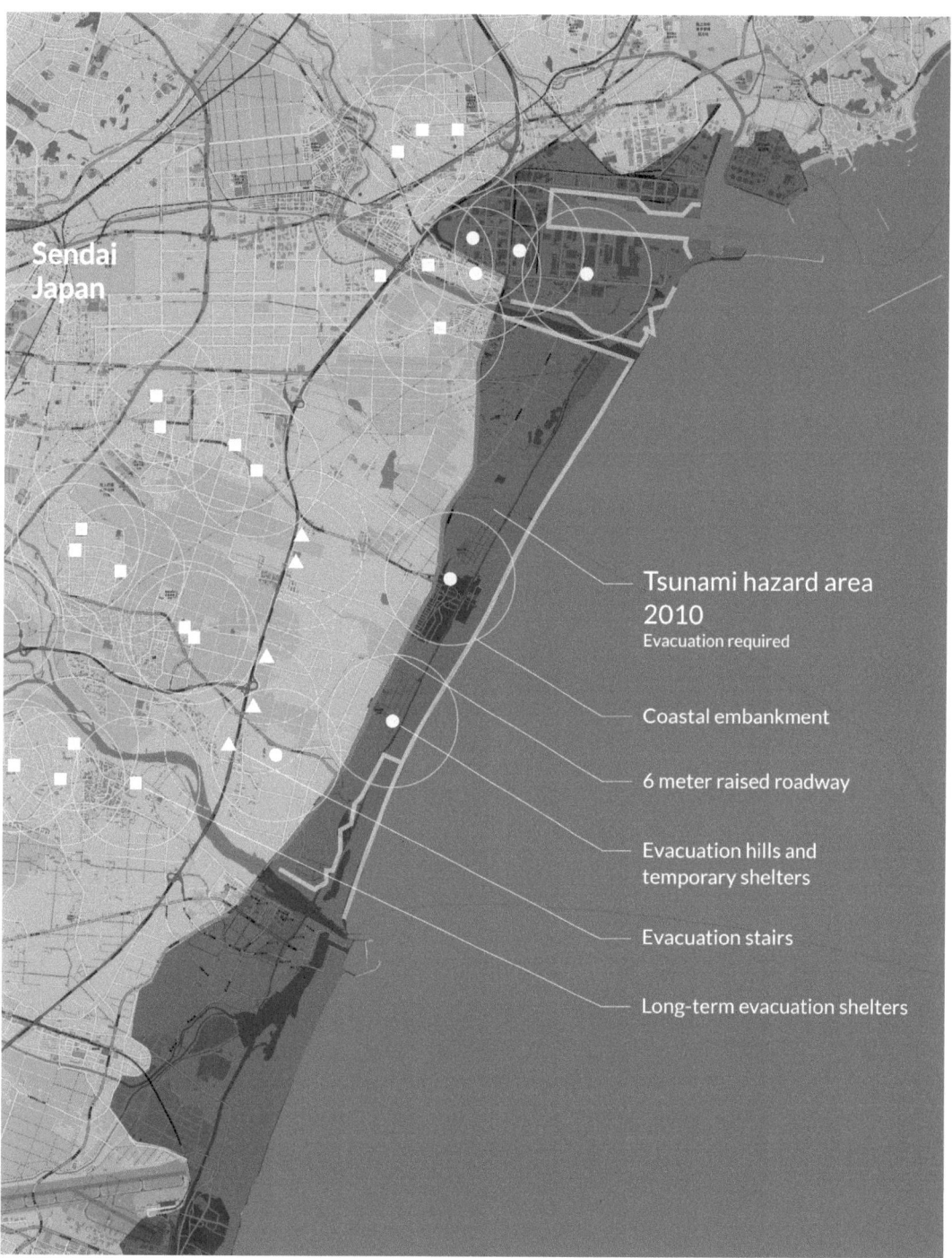

Sendai
Japan

Tsunami hazard area
2010
Evacuation required

Coastal embankment

6 meter raised roadway

Evacuation hills and
temporary shelters

Evacuation stairs

Long-term evacuation shelters

1.1 Tsunami hazard extent modeled prior to 2011 event in Sendai, Japan.

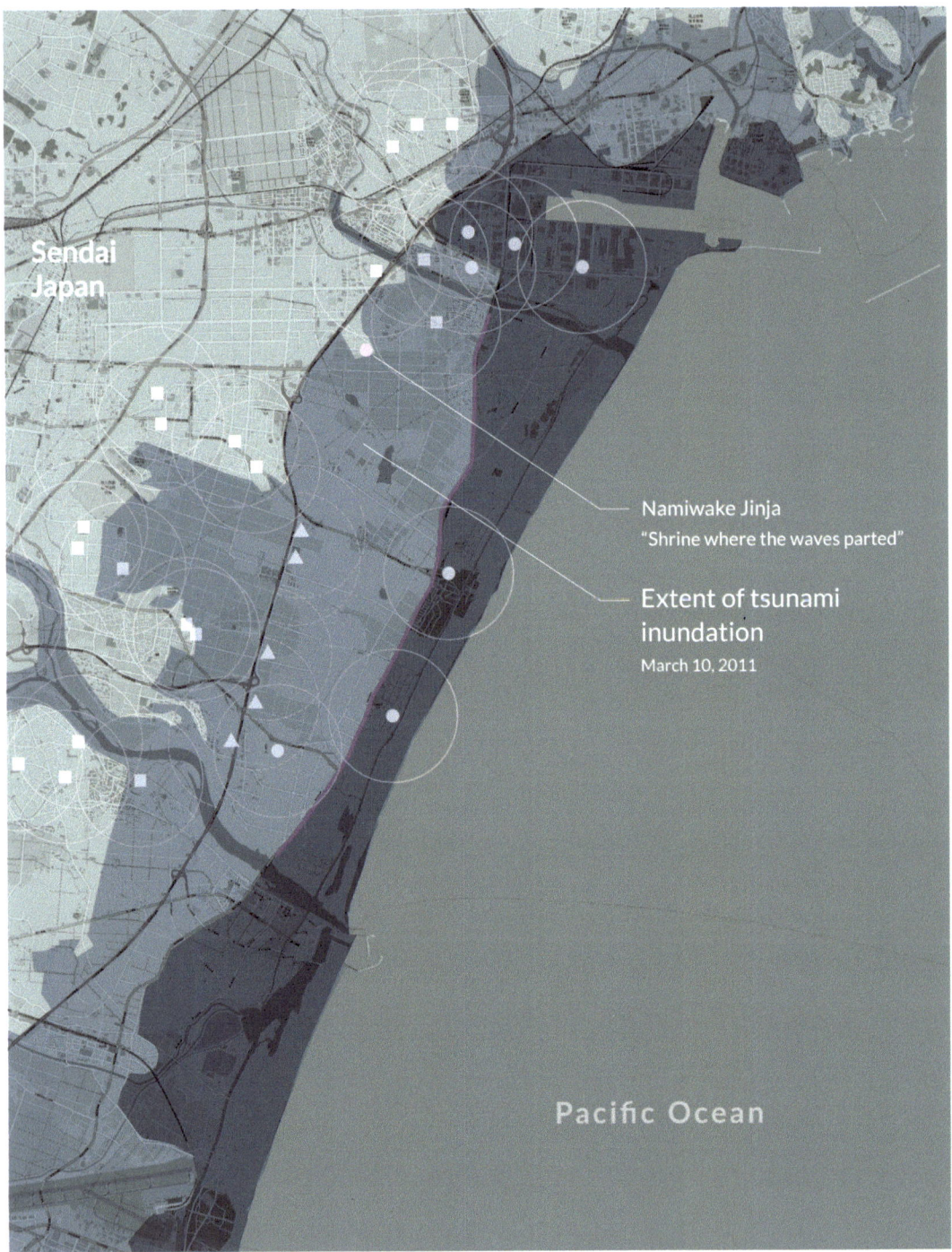

Sendai
Japan

Namiwake Jinja
"Shrine where the waves parted"

Extent of tsunami
inundation
March 10, 2011

Pacific Ocean

1.2 Tsunami impact map and actual extent of 2011 Great Tsunami on Sendai, Japan.

made things worse. Many families failed to evacuate because the time of the tsunami's impact stated in the warning had passed. The warning's inaccuracies led to complacency and loss of life.

The plains of Sendai were not the worst area of damage and loss of life. In the Iwate prefecture north of Miyagashi, the northeastern Tohuku coastline undulates between rocky cliffs and small tributaries that flow to the coast. The valleys that pierce the hills of the coast are called *ria*. Residents are never far from high ground but because of the steepness of the topography and the narrowness of the valleys, waves come in very fast, the *ria* funneling the water's kinetic energy. It is here in these valleys that wave heights assume terrific size. An offshore wave height of eight or nine meters hit the narrow coastal valley of Ofunato and concentrated its energy, resulting in a 30-meter wave funneling up the valley, tearing buildings out at the foundation and pulling people back into the sea to drown.

The rugged coast has a long history with tsunamis – a history recorded in stories, city decision-making and in the landscape. In

1.3 The city of Miyako on Japan's western coast is situated within a *ria* (valley).

1.4 Miyako perspective of tsunami inundation of *ria*.

Aneyoshi, 30 kilometers north of Miyagashi, a 1.2-meter-tall stone sits upright adjacent to the road winding down to the coast. Carved in its face are the words: "High dwellings are the peace and harmony of our descendants. Remember the calamity of the great tsunamis. Do not build any homes below this point." In 1896, the residents of Aneyoshi lived adjacent to the beach to be close to their fishing boats and nets. A tsunami later that year wiped the buildings out, leaving four survivors. After another tsunami in 1933, residents moved up the valley, built houses and placed the stone as a constant reminder of the potential for harm. In 2011, the wave came screaming up the valley but ran out of energy just before reaching the stone. In nearby Wakabayashi, no such stones exist or they were ignored and the waves traveled five miles inland, destroying 2,700 houses.[2] "Each generation builds stone monuments at the highest point of the tsunami that struck their homes, then forgets their lessons, their faded stone lettering a metaphor for collective amnesia."[3]

1.5 A warning stone to mark elevation of the highest tsunami inundation, Aneyoshi, Japan.

In the Sendai plain, the tsunami hazard map underestimated the extent and height of the water, despite sophisticated modeling. Located in a low-lying area, five kilometers west of the shoreline and several kilometers from the line that the 2010 tsunami hazard map indicated would be the tsunami high-water mark, sits a small Buddhist temple, called Namiwake Jinga or "Shrine of the Parting Waters." Built in 1703 and moved to its present site in 1835, the temple warns/protects the Wakabayashi neighborhood. In the 2011 tsunami, water reached the temple but did not damage it, before receding.

Designers must embrace the dynamism of the landscape, whether a literal movement of the earth such as in earthquakes or more gradual shifts of erosion or climate change. Three general methods for accounting for landscape change exist: (1) modeling; (2) analysis of historic patterns; and (3) a dialogue with poetry and stories. The first predicts the future (or attempts to). The second looks to the past. The third, which I return to in the second half of the chapter, combines the past and present. In 2011 Japan, the tsunami dramatically illustrated the limits of tsunami models. It is impossible to document all assumptions, predict severity or timing, and accurately estimate danger. Yet, it is the preferred method of contemporary science because models offer our best *estimate* of the outcome of interactions of natural and cultural processes. Most modeling is beyond the expertise of the architect or planner who rely on others to provide predictions. When designers work along rivers[4] or the coast,[5] we rely on a general knowledge of landscape processes applied to specific terrain, often bringing our spatial expertise in the form of digital terrain modeling or GIS …

… as long as we acknowledge the unpredictable nature of change. Long-term experience serves as a balance, the stone tablets and shrines providing concrete reminders of past events and bolstering accuracy.

Asking local people, reading stones in the landscape turn out to be a valuable practice of landscape dialogue.

LANDSCAPE CHANGE

Landscapes shift. Tectonic plates slip past each other. Rivers spill over their banks and find new channels. Air cools and winds move down the valley at night. If you drill down far enough, rocks flow like liquid plastic.

Landscapes erode. Banks slough. Abandoned buildings crumble. Plant material decays, breaking down into humus. Water pours into a crack in the rock, freezes and the rock splits apart. Venice sinks.

Landscapes accrete. Ash spews from a volcano. Waste from millions of people accumulates in landfills. Buildings arise. Trees grow. Sediment flows down the river to be deposited at its mouth.

Change is not only the slow change over time – eroding banks, sediment deposition, forest succession – but the rapid change of catastrophic events – volcanic, seismic, hurricane, landslide. So how do designers incorporate this change into an analysis of the landscape? Ecology – the study of the relationship between living things and their physical environment – has long dealt with change, wrestling with its trajectory and its frequency. The sub-discipline of landscape ecology embraces landscape change. The spatial component of landscape ecology – things connect – and the temporal component of landscape ecology – things move and change – shape a process-oriented understanding of place.

Landscape ecology is both simple and complex. Its simplicity lies in geometric relationships – of patches, corridors and edges – that inform ecologists' view of the world. The landscape wants to connect; wildlife mating, birds flitting between trees, snakes and gophers expanding their territory. Each species relying on a connected meshwork of plants that form a continuous ecosystem to find mates, food and shelter. Its complexity lies in its incorporation of the ecological disciplines – the patchiness of invisible soils, the intricate movement of birds and the challenge of predicting anything, whether animal behavior, weather, urban development or the next earthquake.

Landscape ecology is so complex that despite ever more complicated models of change, we can only "predict" with a coarse brush. The landscape speaks, but it is hard to hear. It takes just one 9.0 scale earthquake off the coast of Japan to show us our limitations. Turner and

Gardner ask why it is difficult to explain and predict landscape change. They offer four potential challenges:[6]

1. Multi-variate interacting drivers – an unwieldy phrase indicating the complexity of landscape relations – ecological factors, such as weather, that influence other factors in a myriad of ways. The tremendous potential exists for modeling to leave something out, misjudge the consequences of interactions and/or falsely simplify the interactions of multiple factors.
2. Thresholds and non-linearities – small changes in the landscape that cumulatively result in large or catastrophic shifts. Once a threshold is reached, the levee is breached, the sea wall fails, the species migrates or becomes extinct.
3. Social-ecological systems – an acknowledgement that humans "disturb" the complex workings of ecological processes. (And the corresponding attitude among some ecologists that if we did not have people around we could figure these things out!)

patch size decreasing over time

  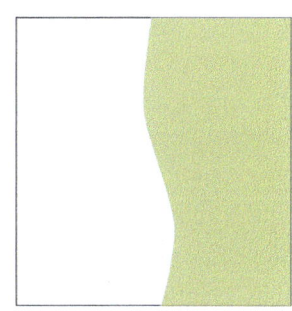

simplification of edge conditions over time

1.6 Patterns and processes of landscape ecology acknowledging temporal change.

4. Limited ability to perform experiments – a recognition that the multiplicity of landscape factors interacting precludes using experimental approaches where it is necessary to isolate variables and establish cause and effect.

While landscape ecology offers a framework for understanding many of the processes and patterns of both space and time, it is constantly seeking better, more accurate ways to establish the why, the causes of these changes, to better be able to predict how landscapes might change.

SITE SPEAKS BUT SAYS ONE THING

Landscape ecology and its focus on change confronts the systemic work of bounding, recording and maintaining property. The word "site" captures this idea of a bounded piece of land. It conveys a static quality both spatially and temporally. A site is a piece or portion of the landscape, a useful way of organizing the world. Cities can designate specific parcels to provide urban amenities and necessary infrastructure. Homeowners can own property that provides a living space for their family. Farmers can plant crops each year on a certain amount of acreage. Real estate investors can buy, sell and build on properties. Site is geographic specificity. It is working within limits.

Many designers recognize the distinction between landscape and site intuitively, yet frame the landscape as exclusive, bounded property written not on the land itself but on pieces of paper at the County Assessor's office. The framing of site informs how we practice design and architecture. It informs how we visit a place, how we analyze its properties and how we communicate with others about what should be done.

The dynamics of the landscape enforce their own reality, whether that reality is in the daily movement of energy, wind and creatures or a catastrophic "external" event. Site does not hold together as a separate whole. Site then, as opposed to landscape, situates within a system of fragmented, codified and controlled spaces which delimit the boundaries of practice. Despite the best of intentions, design-within-the-lines becomes analysis-within-the-lines. And, as we have learned from landscape ecology, analyzing within lines is fatal to an understanding of landscape and its processes. Site boundaries cannot prevent change, movement or the passage of time.

LISTENING TO CHANGE

It is difficult for humans to perceive (slow) changes and the potential for change in the landscape. From building in a flood plain to the erosion of agricultural soil, change eventually surprises us. The surprise may arise from an inability to predict the future, a lack of will to address the potential for change, or an unwillingness to associate change in the landscape with our own change as people. In one sense, to face change is to face mortality and the death of that which is around us, to confront the tsunami wave crashing over the sea wall as well as to confront the inevitable erosion of our bodies. Humans would like things to either stay the same or grow and progress. There is no room for decay.

To break through the fear of change, to force a confrontation, we require stone tablets embedded in the ground at the limits of the cresting wave. We tell and retell myths and stories to remind us of the tenuous nature of life. We do this through markers but also myths, relationships and spirits, all to explain the vagaries of life. In Japan, mythic creatures, *yokai* (妖怪), inhabit the mountains and streams and coasts of the islands. Strange spiritual beings of Japanese folklore, the *yokai* interact with humans in positive and negative ways to help explicate natural processes in the landscape and the cultural consequences of ignoring change. They are supernatural manifestations of the unknown or misunderstood. *Yokai* remind us to appreciate the world. They occupy a special place in Shintoism, woven into the fabric of Japanese society – the *Tengdu* in a river or the *zashiki-warashi* in a house. One such *yokai*, Namazu, a giant catfish, resides on the ocean floor.[7] Usually constrained by the thunder god Kashima, Namazu constantly strains at its tethers. When it breaks free, the earth shifts, water expands and a giant wave moves towards land.

We require words in stone, metaphors of our frailty and offerings to the local gods.

Landscape ecology may address change, but gaps remain. Gaps in predictive models that measure the frequency and likelihood of events. Gaps in infrastructure investment to perpetuate the uneven nature of climate change impacts. Gaps in a city's tree cover signaling the location of a Latino community who experience higher summer temperatures.[8] The next wave of landscape ecology may not only embrace the dynamics of the landscape with its emphasis on habitat/living and corridors/connections but fill in gaps of equity and climate.

1.7 Nineteenth-century woodcut of Namazu, the catfish, being brought under control by the god Kashima while yonaoshi dimyojin provides money to the poor (who take advantage of the tsunami's destruction).

LANDSCAPE DYNAMISM

How does the landscape speak? And how are we, as its inhabitants, to listen and understand? It speaks through movement, through flows and the accumulation and deaccumulation of materials and energy. It speaks abruptly through tsunami warning sirens and gradually through the accumulation of detritus on the forest floor.

Landscape dynamism is the changing nature of both landscape and our relationship within it. It has several qualities: (1) the motion of material and energy; (2) interactions between objects and flows; and (3) a framing of the world as always in motion. In ecological terms, processes create patterns, patterns which themselves shift over time. These patterns both reflect and reinforce cultural norms.[9] In a narrow, ecological view of landscape, landscape dynamics are mechanistic flows that can be modeled to assess and predict landscape change. In a more robust view of landscape, landscape dynamism incorporates the gestalt of continuous (and mysterious) change with all of its complications and surprises.[10]

Historian John Stilgoe, in his book *What is Landscape?*, explores the etymology of earth-bound words: hill and dale, berm and swale. For the term *landscape*, he turns to nineteenth-century Flemish sailors and fishermen, rather than the German nobility's *landschaft* which contains an internal division of natural and cultural spaces.[11] Sailors on the Frisian coast of the Netherlands used landscape to describe "the land thrown up against the sea." They patched dikes, built windmills and pumped water. Stilgoe likens it to a child with a pail and shovel constructing a sandcastle. This suggests landscape is not something we observe, but something we create, construct, maintain and bolster, and it in turn maintains us. We do not work with static stuff. Landscape dynamism is not just the result of geologic forces of millions of years, occasionally wreaking havoc on the earth, but the daily interactions of people, land, water and air.

How then do we account for change in the landscape? As people move through space, as the earth erodes and water flows, what aspects of our field and office work might embrace change and even anticipate change so that our designs can respond? Listening and speaking.

First, designers identify the most relevant time period. Fluvial geomorphologists have found that rivers and streams are shaped, not by the small daily flows of water that lack the power of moving sediment and not by the catastrophic flood events which are too infrequent to shape a river's banks, but by the two-year flood event – frequent enough to maintain a river's shape, while powerful enough to make a difference.[12] In the same way, landscape dynamism may assert itself in the less observed, annual and biannual events that shape space. The University of California, Davis, where I work, has a central quad like many other universities – a large rectangular patch of grass in the middle of campus used for daily meet and greets, lunches, Frisbee and advertising student organizations. However, it is the annual Whole Earth Festival, happening every spring, that shapes this space. Hundreds of booths selling crafts and food, drum circles, people walking and dancing and singing. Tens of thousands of people trample the grass for three days. To analyze the quad at U.C. Davis, the designer could observe how people use the quad on a random school day in October. This would be important information, but without understanding the annual Whole Earth Festival event and its impact on the soils, the grass, the roots of the enormous cork oaks and deodar cedars that frame the quad, the designer's time would be wasted. The grass and soils of the quad must be able to accommodate the heavy foot traffic of the festival, not the students lounging at lunch.

1.8 The shaping of dikes and follies in the Netherlands.

Second, designers circumscribe the space of change. Where do change events arise? From how far away? Landscape dynamism embraces diverse spatial scales as well as temporal scales. Limiting analysis of change to a site or the limits of a property boundary would subject the process to the same dangers the Japanese experienced in the tsunami of 2011, as many natural and cultural processes are not subject to our boundaries. Deforestation upstream, for instance, leads to flashier surface runoff flowing downstream and increased flooding. If the amount and change in impervious surface in a watershed is not considered, designers might miss a key driver of hydrologic change on the site.

CASE STUDY: SEASIDE, OREGON

During the Great Earthquake of Eastern Japan, the tsunami fanned out in all directions across the Pacific. Three hours after the earthquake a wave three meters high washed over the coast of the Aleutian Islands in Alaska. In the middle of the night, the city manager of Seaside, Oregon in the United States received notice of the earthquake and tsunami. He alerted authorities, initiated the first level of alerts and waited for it to reach their shores. At 7:46 am, ten hours after the earthquake, a two-meter wave moved up the beach of Seaside, touching the top of the beach revetments, before sloughing back into the ocean. It lacked the drama of the Japanese tsunami but illustrated the interconnectedness of the Pacific "Ring of Fire." Waves are not mono-directional. When water is displaced, as a pebble tossed in a pond, the ripples radiate outwards. A 9.0 earthquake and resulting tsunami in 1960 off the coast of Chile moved across the ocean at a rate of 600 km per hour, before reaching Japan unexpectedly (they had felt no earthquake) and causing damage to 1,600 homes and killing 138 people.

Seaside, Oregon sits along a complex estuary on the Pacific coast on the eastern edge of the Ring of Fire. It is difficult to think of a more dynamic landscape. Not one, but three rivers flow into the ocean at this point, one from the north and two from the south before meeting and abruptly turning into the ocean. The town sits on shifting ground. An earthquake along the Cascade Subduction Zone off the coast would trigger a tsunami that would reach Seaside in 15 to 20 minutes, lapping up the beach or wiping out the whole town, depending on the earthquake's proximity to the Oregon coast and the height of the wave's crest.

Yet the town of 6,000 people, like any town, projects a permanence drawn from local history and economies. Or that is the intent. Chamber

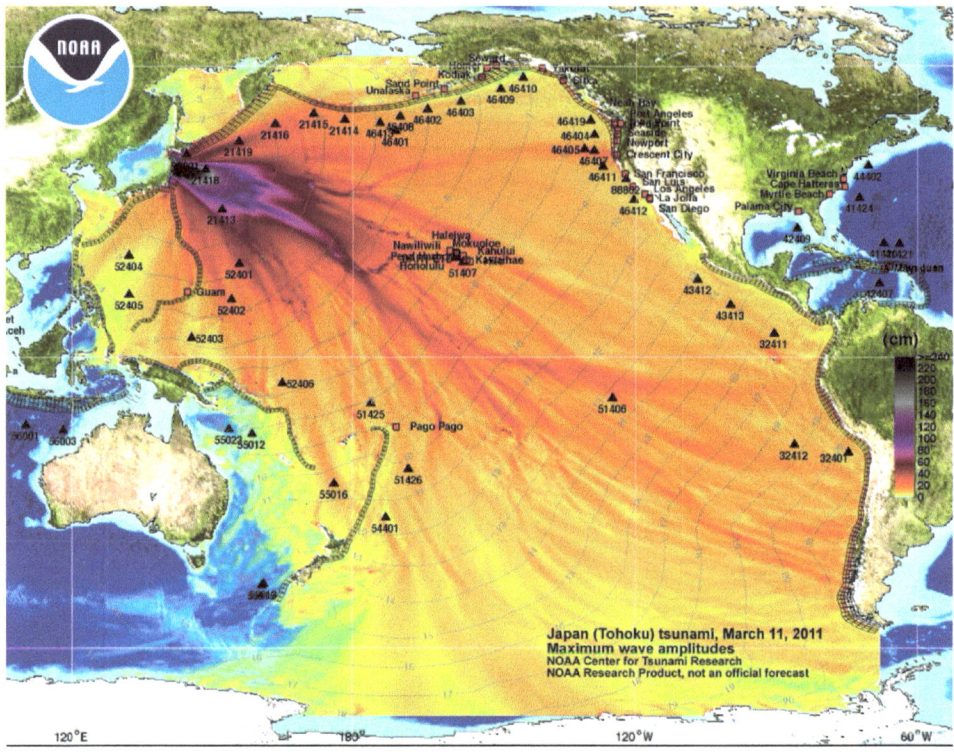

1.9 Tsunami wave amplitudes for March 11, 2011 from Japan calculated with the MOST tsunami forecast model.

1.10 Aerial view of Seaside, Oregon and the confluence of the Necanicum River, Neawanna Creek and Neacoxie Creek.

of commerce and tourism brochures link the town to the western-most point reached by Lewis and Clark's expedition just to the north. It seems obligatory to mention this when discussing the town, along with nostalgia for the lumber industry, a series of mills and ponds in the southeast. The focus appears to be "we have been here a long time."

Developers have proposed building a residential neighborhood here on these shifting sands ("a house built on sand …"), but most residents recognize the tenuous nature of their place. However, the town's true purpose, reflected in its name, is to get people to the ocean. The beach is wide and long and sandy (some sand is imported). A boardwalk runs along its upper edge past salt taffy shops, arcade games, stores selling kites and wind chimes. The openness of the beach, the straightness of the boardwalk and the proximity of it all to commerce give Seaside a certain obvious quality common to beach towns from Santa Monica or Atlantic City in the United States to Brighton, England. Visitors come from the Portland urban area with disposable income in search of a particular ocean experience of the beach.

The beach emphasis in Seaside is a blessing and a curse. The beach setting brings in tourist money in the summer. In the winter, it can seem desolate. The setting, the narrow estuary between ocean and hills, prevents additional development as well as alternative economic streams during winter. The town sought assistance in the development of a master plan to open up the rest of the town, to highlight Seaside's "backyard," the rivers and the forested slopes behind the town.

Prior to a community master plan, architects from our firm had visited the city and met with people in the town to discuss the place. The city desired a plan to expand recreation options from the beach to the rest of

1.11 Geologist at Necanicum Estuary explaining shifting sands.

the city. At this preliminary stage, before a project has started, before a contract has been signed, the analysis of a landscape's dynamism may be fairly limited – walking around, talking to people and taking pictures. As designers of space, we prod, assess, propose ideas, shoot them down. In conversations with city staff, we developed some preliminary goals that carried us through the interviews, contract development and initial stages of the project with remarkably few changes: (1) account for landscape dynamism; (2) shift attention to the interior landscapes; and (3) interpret local history. The goals run parallel, compatible visions arising from the situated nature of the rivers and forests.

The first planning goal was to work with the landscape dynamism of the town. The reality of Seaside, Oregon means any improvement to open space needs to also improve emergency evacuation – to open bigger conduits for people to get to high ground. The Pacific Northwest lacks an early warning system such as found in Japan. When the big one comes to the Cascade Subduction Zone, the compression waves will radiate outward but instead of sensors registering initial waves, the second, more damaging waves of the earthquake will hit without warning. Those in Seaside who survive the earthquake will have a few minutes to run east, uphill to higher ground.[13] There will be no time to gather a few meaningful things, tend to broken bones or rescue others. Designing in Seaside begins with tsunami inundation maps of which there are many (see Figure 1.14). The 1964 Anchorage earthquake sent a tsunami down the Pacific coast; the line of inundation from this tsunami is clearly marked in maps of Seaside. Further back in time, historical records and oral stories from the Chinook and Clatsop tribes contribute to estimates of potential inundation. Designing begins and ends with tsunami routes; how to get someone from the beach to the hills in less than 15 minutes. (If the Juan de Fuca plate moves like the Honshu plate in 2011, they will have even less time.) And it is this potential movement – during a catastrophe but also during the busy tourist summer – that motivates city staff and residents.

The second planning goal was to inspire visitors and residents to move past the beach into the interior. Nothing would completely replace the beach as a draw for a town named Seaside but the local ecology and history of the interior offered activities of interest to people. A kayak rental place perched above the river. Residents frequented sports fields, trails for hiking and the edge of an intact wetlands for wildlife preservation. Of course, city residents did not necessarily agree on these

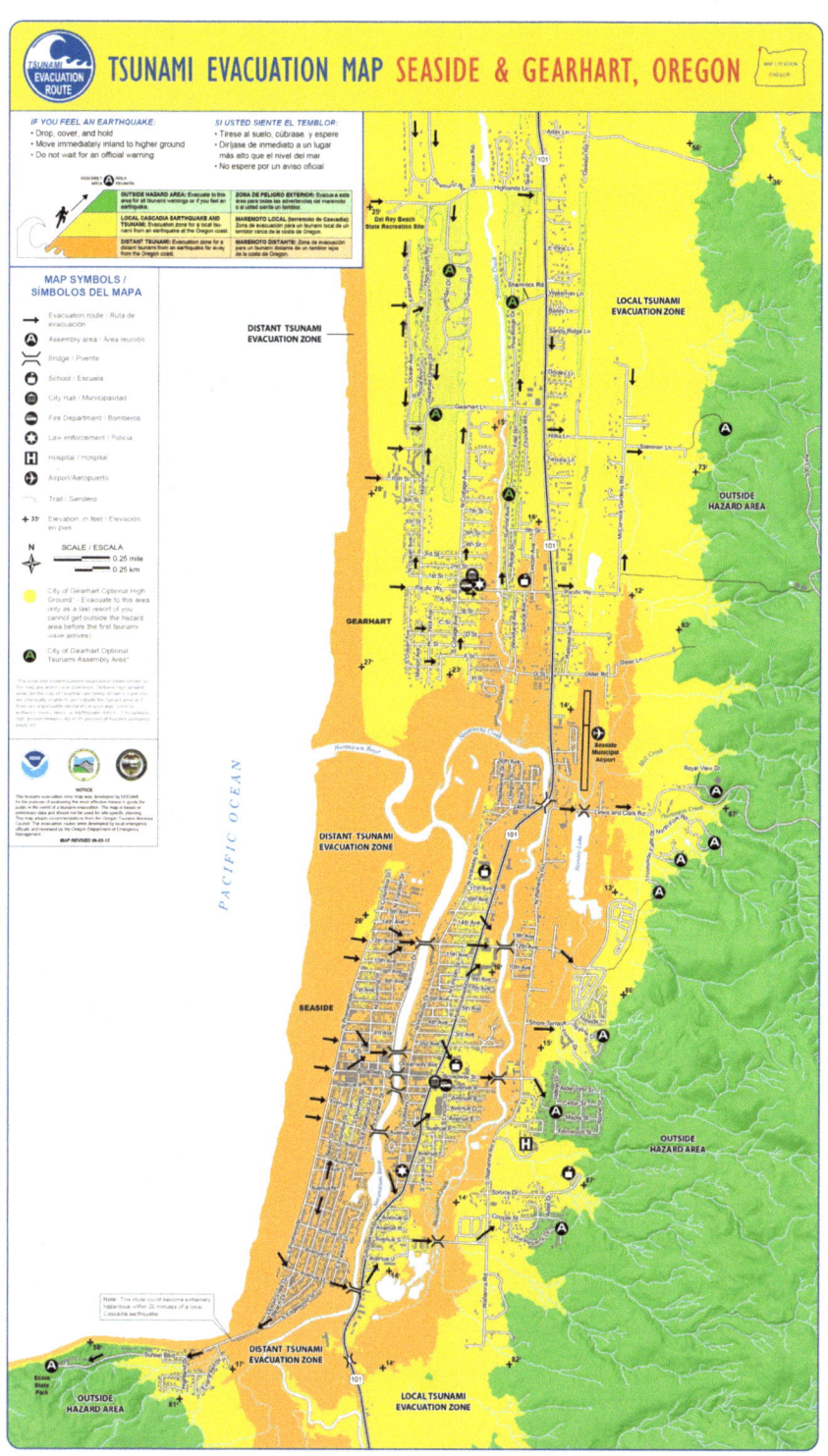

1.12 Tsunami evacuation zones and routes for Seaside, Oregon. Orange indicates evacuation in response to a distant tsunami. Yellow indicates evacuation in response to an earthquake off the Cascadia coast (Oregon Department of Geology and Mineral Industries).

landscape activities. In one public meeting, I discussed the potential for a jogging trail running along the toe of the forested slopes to the east of the city with a member of the parks board. Another resident active in an environmental non-profit group came up to me afterwards, aghast at what she had overheard. Were we really considering having people jog through these sensitive wildlands? We had to work closely with both recreationist and wildlife preservationists to provide spaces for people and animals.

The third goal was to present a local history for interpretation. Immediately, we were faced with a challenge: Whose history? One history of Seaside is already told in books and magazines. It is told on the tourist bureau's website. And it goes like this … Once upon a time, Lewis and Clark arrived at the end of their transcontinental journey across the United States. They made friends with the local Clatsop and Tillamook Indians and set up a salt works near the beach. Time goes by. Trappers, then foresters, wandered the coast. A railroad was built from Portland to Seaside, then a hotel to accommodate the sudden surge of tourism. Always this relationship between the waves and the shore pulled people to the place. That story is, at best, incomplete and, at worst, has removed whole peoples and time periods from the landscape of Seaside to establish a normative view of who is a citizen of Seaside and who is not.

Many in Seaside recognized this. The effort to define local histories accompanied the effort to shift focus from the beachhead of Lewis and Clark to the interior trade routes of the Clatsop and Tillamook peoples' (continued) presence. Before contact with the West, a small but robust network of tribes occupied the inlets of the Pacific Coast. The Clatsop, "fish-eaters," centered their villages on a point above Astoria. The Tillamook ranged along the coast to the south. Chehalis lived around the mouth of the Columbia River. Landscape stories of the past, whether on a tourist bureau website or a map of Indian "territory," may not capture the ingenuity and dynamism of the tribes in response to white settlement. People of all three tribes ranged up and down the coast, into the interior, along the rivers to fish and move and celebrate. In the Clatsop origin story, the original peoples of the area faced starvation. Talapus, the great wolf-spirit, gave them salmon as a gift. The Clatsop honor Talapus by treating the first salmon of the year with respect, cutting it only lengthwise and sending its bones back to the water for renewal.[14]

After disease decimated the population, the remaining Clatsop were expelled from the Astoria area. Rather than "vanish," a significant group of them moved to Seaside, families occupying the interior of the town, practicing

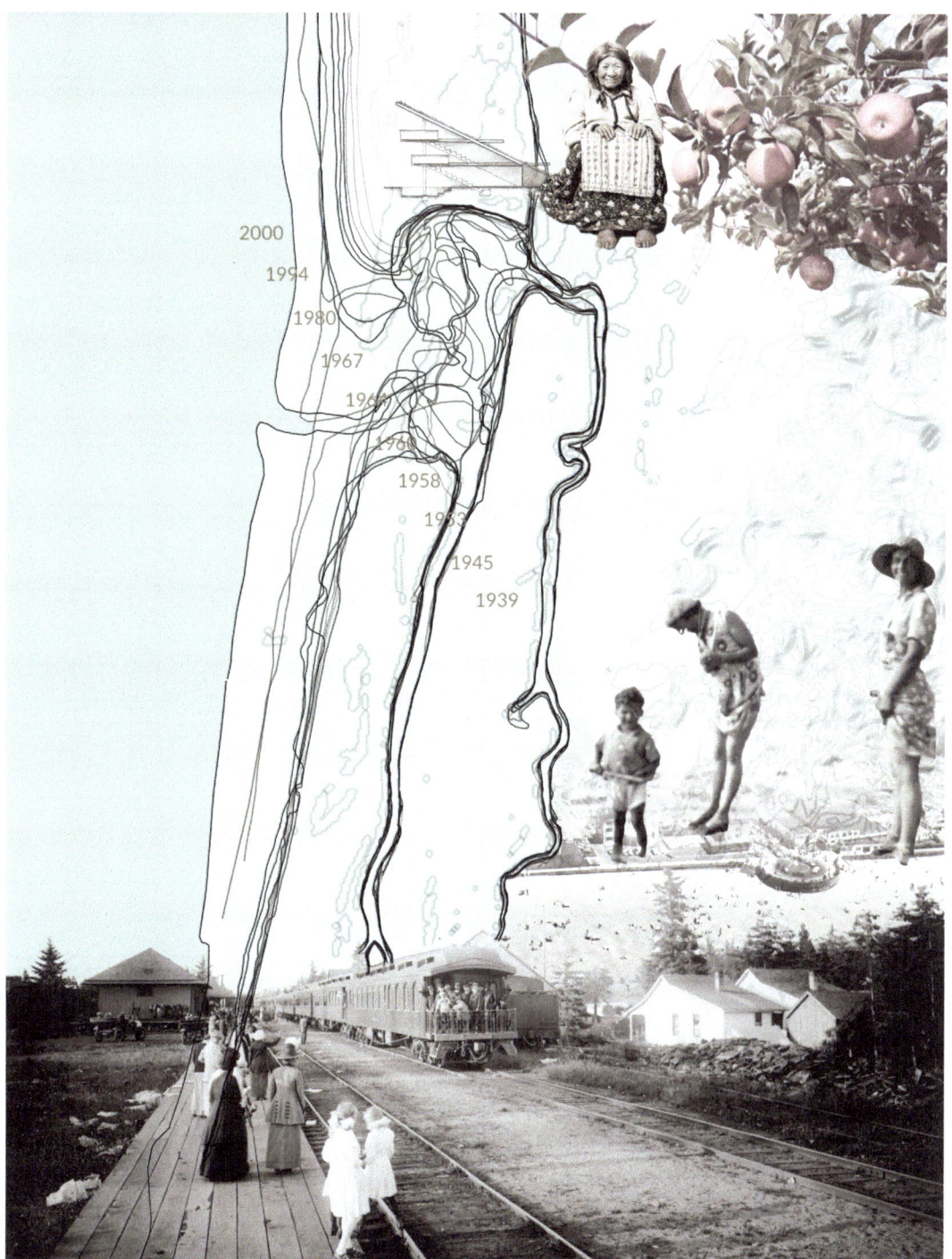

2000
1994
1980
1967
196
196
1958
196
1945
1939

1.13 Seaside as dynamic landscape of Clatsop, tourists and ocean.

traditional crafts of basket-weaving, fruit production and canoe building. In this period, Seaside was known as "Indian Place" due to the prominence of Clatsop families.[15] They navigated the slow creep of materialism, the initiation of tourism with the advent of the Seaside Hotel and changes in traditional ways of life. Many residents wanted to acknowledge and re-orient the town to a more comprehensive, more multi-cultural narrative.

This history and contemporary culture cannot be "seen" when walking around the landscape, when initially recording observations. Even if one came upon the fruit orchards planted by Clatsop families in the interior of the town, one would not necessarily have knowledge of where they came from and who planted them. But the designer practicing landscape dialogue would know what kinds of questions to ask, reading histories before visiting the site so one's questions can be asked of the landscape itself – to go looking for the orchard that sustained families and led to the selling of fruit outside the hotel, along with the telling

1.14 Abandoned orchards along the Necanicum River in the interior of the town of Seaside.

of stories to tourists interested in "local color." A landscape dynamism approach incorporates the socio-ecological changes of peoples and their movements – whether up and down the coast or back and forth between the coast and the sea.

In the end, our team of designers proposed a three path/level plan – beach, river, hillside – that would integrate human and animal movement through and within the town. At the north and south termini of the paths, vertical cultural centers would raise the visitor up to see the

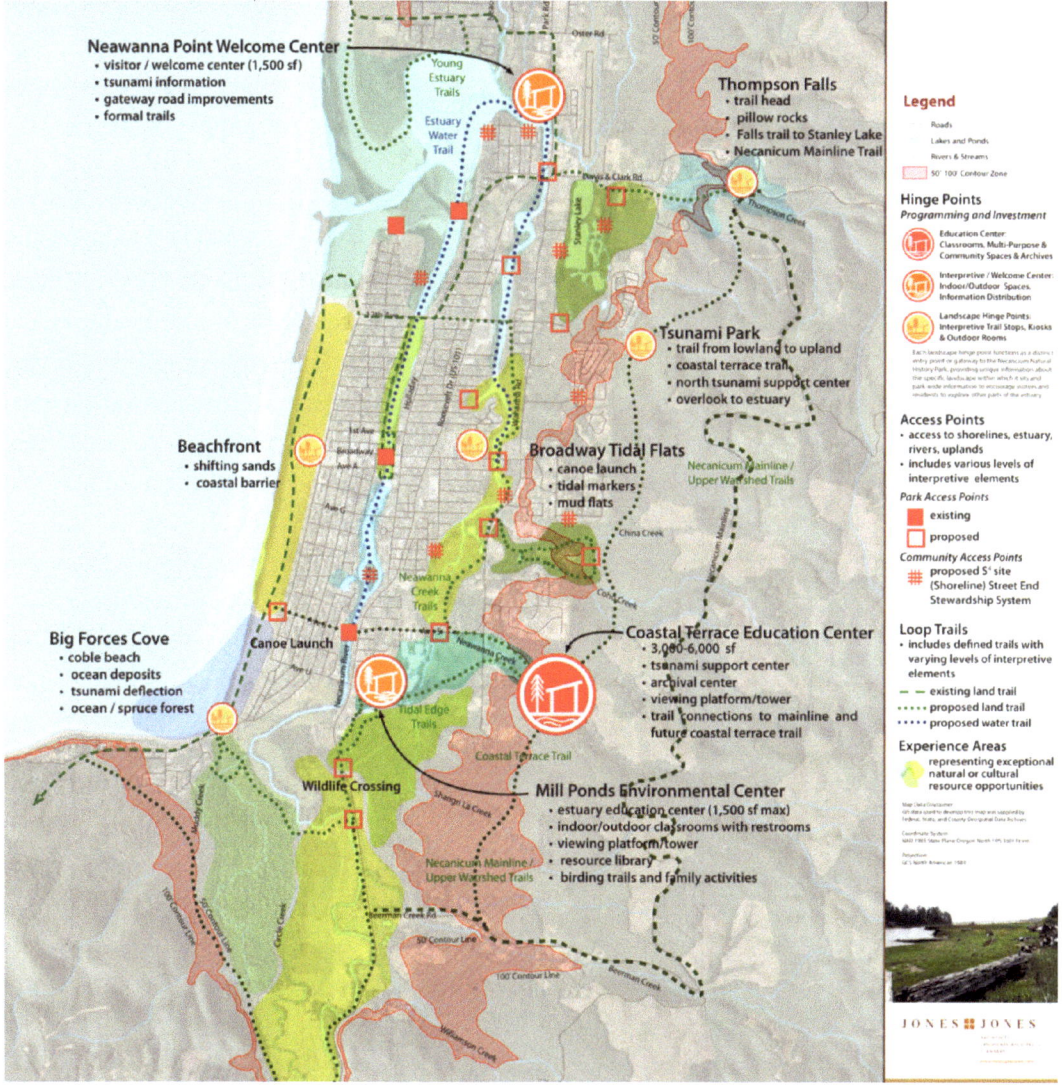

1.15 Seaside open space and recreation plan, City of Seaside by Jones & Jones.

sweep of estuary. The southern end terminates in the wetlands critical to wildlife, freshwater ecology and the Clatsops' cultural engagement with this place. Tourists will better understand the tenuous nature of the cultures that dwell here. Landscape dynamism is a recognition and evaluation of the changes that have occurred and will occur again in the future, an acknowledgement of the need to increase economic and physical resiliency, and a focus on people immersing themselves in the whole landscape, not just the beach.

PRAXIS: MAPPING FLOWS

To understand landscape dynamism, map the flows, not the static elements. Mapping flows anticipates change based on what happened in the past. It means generating new, more plastic ways of looking at landscape. This contrasts with mapping suitability, a means of layering static data on top of each other to generate a suitability map for development or conservation. While the McHargian approach to evaluating the suitability of greenfields for their appropriateness for development remains a useful method of analysis, the designer needs to engage with the dynamic flows of the landscape.[16] A proposed development location and configuration may not be able to accommodate the shifting soils and slopes, the increased impervious surfaces of the watershed or neighboring gentrification, if it relies on static maps.

Mapping flows, unlike alternative futures mapping or scenario planning, does not so much predict future outcomes as wrestle with ongoing changes. This understanding itself is always partial and fluid but can drive design forward in positive ways. Mapping flows acknowledges three processes of analysis:

1. Understanding dynamic processes.
2. Capturing historic change in cartography.
3. Animating these processes in technical map sequences.

While acknowledging the messiness of the process, mapping flows starts with an inventory of dynamic processes, followed by prioritizing those with the greatest impact. This means that in Seaside, Oregon, despite the absence of a major earthquake in the last 300 years (or really *because* of the absence of a major earthquake in the last 300 years), the vagaries of the Juan de Fuca plate and the ocean's response are paramount. In this prioritization process, let temporal scale assist. Landscape ecology

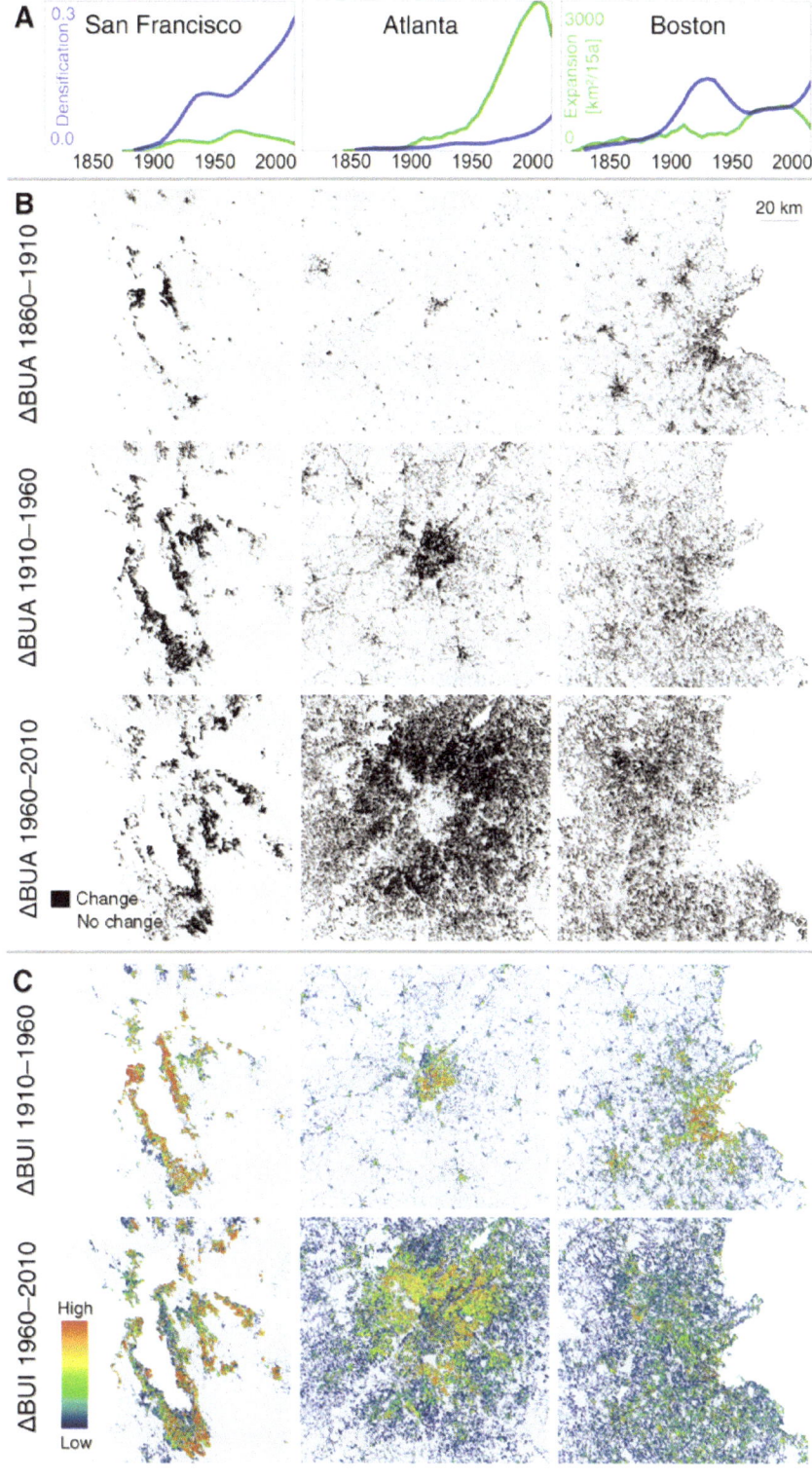

1.16 Long-term change of Built-Up Area (BUA) in three cities in the United States: San Francisco, Atlanta, Boston.[17]

classifies change using two scales of time: (1) the seasonal (intra-annual) and (2) long-term (inter-annual) change. For the purpose of designing buildings, infrastructure or landscapes, I add two more: (3) daily change and (4) event change. In order of priority in planning:

1. Event change is not on a continuous time scale but consists of sudden events that influence a place. Flooding, earthquakes, famine or war dramatically disturb the landscape.
2. Long-term change ranges from the geologic change of continental drift to contemporary climate change as the earth warms. Population flows, such as immigration from rural areas to cities, are a good example of long-term change.
3. Seasonal change occurs within the annual cycle. Winter, spring, summer, fall – a changing climate triggers plant growth, rain or snow, students home from school and the migration of animals. The classic example of the midwestern farmhouse planting trees along its northern edge to shield the house and crops from winter winds. But the reverse is true … the Australian farmhouse shielded from the hot summer winds.

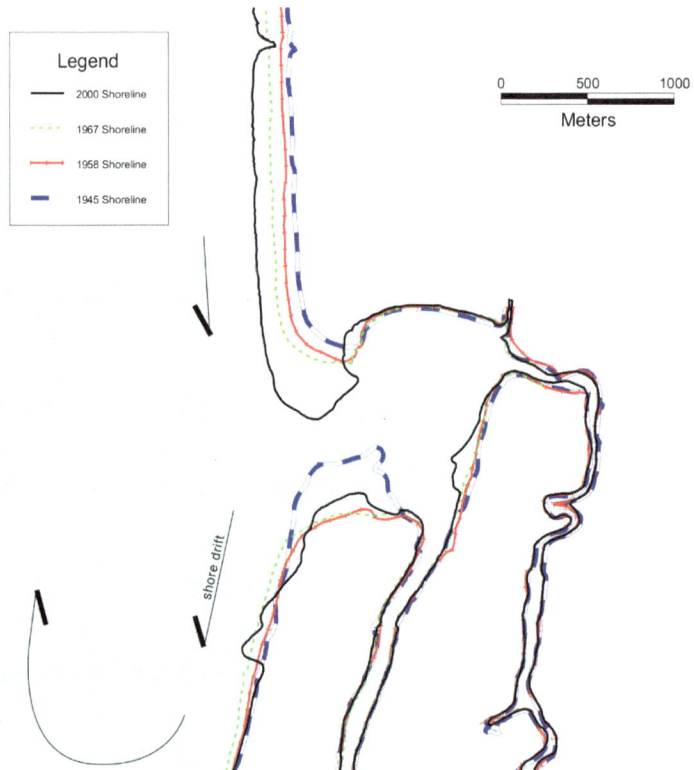

1.17 Change in Seaside's shoreline over time, from 1945 to 2000. Arrows show direction of littoral sand transport.[18]

4. Daily change includes the daily movements of people, water, wind and nutrients through and within a place. This is landscape dynamism at the personal scale, often limited spatially. This is the scale of the habitual, the taken-for-granted use of space.

For each landscape, choose an important event-based process along with a seasonal or daily flow. Choose the most significant events/flows – flooding, tsunami, congregating people, electricity – and map them at the most relevant time scale. The prioritization of flow importance will vary depending on location. For a riverwalk, event flows of flooding compete with daily flows of commuters and recreation for the attention of designers. In a desert town in North Africa, wind and access to water shape decisions in design and planning. In Hong Kong, the flow of energy and the flow of traffic impact people's use of space. People walk, run, drive and ride. The choice is important … we assess the *whole* landscape, so any reduction of the landscape to two flows will only be a starting point for dialogue. Holistic means embracing different times, the changing landscape, as well as different landscape elements. Designers can move past the reductionism of site analysis by choosing a flow influenced by other landscape processes, even a consequence of multiple processes. For instance, in the Seaside area, shore drift moves sand southwards along the coast, eroding some beaches and depositing sand on others. Yet the map of flows over time would also show sand arriving by truck to be deposited on the beach; maps would illustrate oceanic and economic processes by looking at one flow, the accrual and depletion of sand.

Map processes using a time sequence. For infrequent events, mapping could be as simple as a before and after images (i.e. Figures 1.1 and 1.2). For seasonal flows, a sequence might include quarterly maps of seasonal patterns, such as water levels or tree leaf-out. For daily flows, the location and direction of people can be shown as a symbol on a map of movement. In a behavior map, while the location and activity of each person is useful, field notes about where people are coming from and where they are going might lead to better understanding of flow. Each map sequence should provide as many snapshots or still images as possible to illustrate the landscape's dynamism. Soil erosion, circulation, shadows, plant growth and wind can be informally mapped in the field, while long-term processes of encroaching development, invasive species colonization and river migration can be mapped using aerial photography or infrared photogrammetry.

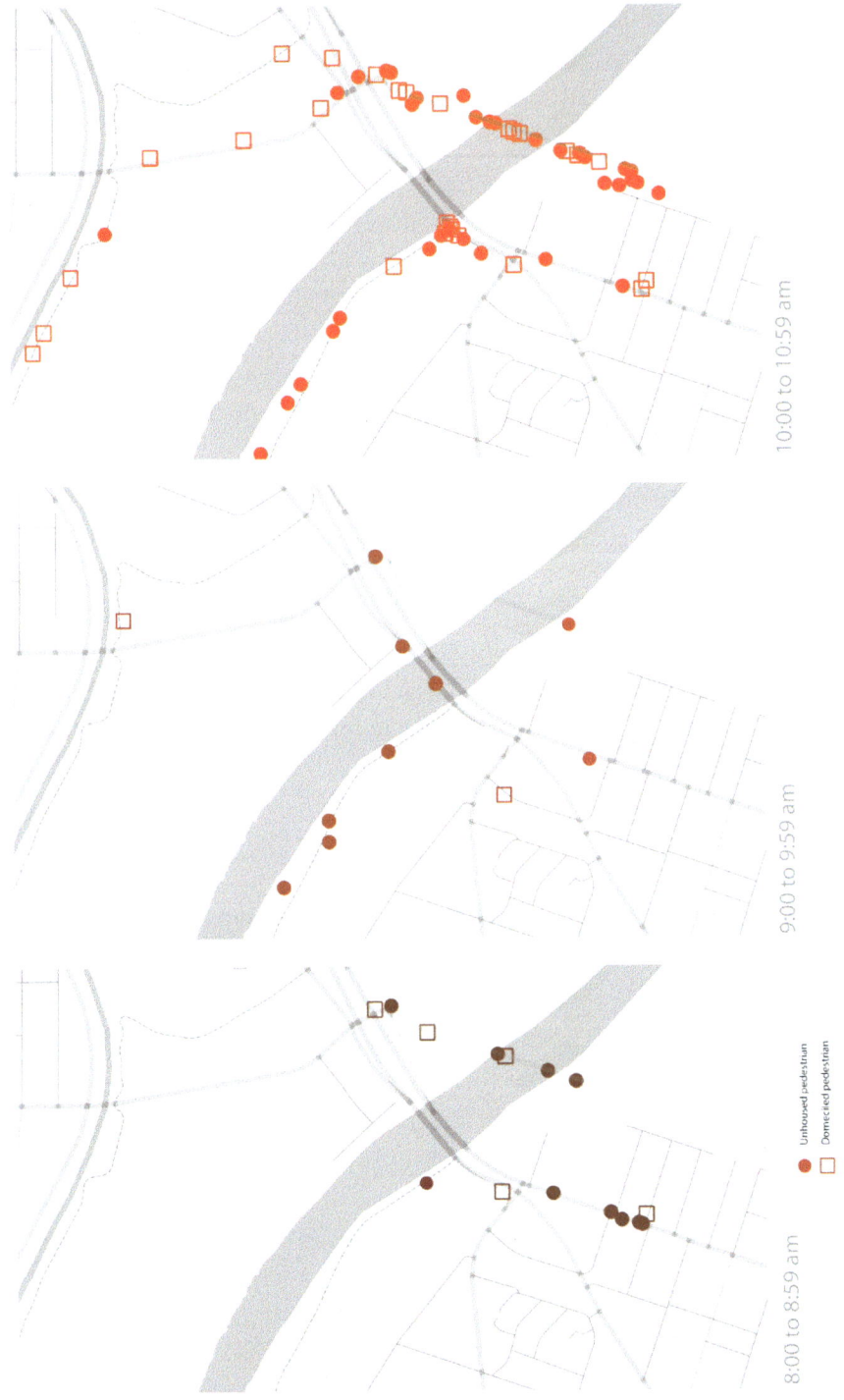

1.18 Snapshots from animated time series of encounters with homeless and domiciled pedestrians in Sacramento.

More dynamic than sequential static maps, GIS animation more closely approximates the flow of data across a landscape. Often, time is condensed to show change over a set period of hours, days or years. Maps that contain animation in GIS show past to present movement as shifting points or heat maps. The following process outlines a general approach to creating an animated map in QGIS; the steps are similar for ESRI's ArcGIS. Rely on online tutorials for a detailed step-by-step process.

1. Add a data layer with an attribute of date and time.
2. Navigate to the temporal tab on the layer information page and enable temporal data.
3. Activate animation controls.
4. Choose a date and time range.
5. Play the animation sequence; adjust the rate of playback.
6. Label the animation to show a changing date during animation.
7. Export the animation as a series of. png files to create a GIF, if needed.

Mapping flows acknowledges change over time, something people can miss in initial studies of the landscape. Each flow weaves with other flows; each process influences others. Landscape dialogue initiates a conversation with movement.

NOTES

1. Accounts of the 2011 tsunami in Japan come from Lucy Birmingham and David MacNeill's book *Strong in the Rain* (2014), as well as data from the Japanese Meteorological Agency and the United States' NOAA.
2. Lucy Birmingham and David MacNeill, *Strong in the Rain*, Reprint edition (Basingstoke: Griffin, 2014).
3. Birmingham and MacNeill, *Strong in the Rain*.
4. Tibor Erős and Winsor H. Lowe, "The Landscape Ecology of Rivers: From Patch-Based to Spatial Network Analyses," *Current Landscape Ecology Reports* 4, no. 4 (December 1, 2019): 103–112, https://doi.org/10.1007/s40823-019-00044-6.
5. Trine Bekkby, Lars Erikstad, Vegar Bakkestuen and Arne Bjørge, "A Landscape Ecological Approach to Coastal Zone Applications," *Sarsia* 87, no. 5 (October 1, 2002): 396–408, https://doi.org/10.1080/0036482021000155845.
6. Monica G. Turner and Robert H. Gardner, *Landscape Ecology in Theory and Practice: Pattern and Process*, 2nd edition (New York, NY: Springer, 2015).
7. Phillip Kennedy, "When Giant Catfish Shook the Earth: The Namazu-e Prints," *Illustration Chronicles* (blog), October 18, 2016, https://illustrationchronicles.com/when-giant-catfish-shook-the-earth-the-namazu-e-prints.

8. Sharon L. Harlan, Anthony J. Brazel, G. Darrel Jenerette, Nancy S. Jones, Larissa Larsen, Lela Prashad and William L. Stefanov, "In the Shade of Affluence: The Inequitable Distribution of the Urban Heat Island," in *Equity and the Environment*, ed. Robert C. Wilkinson and William R. Freudenburg (Leeds: Emerald Group Publishing Limited, 2007), 173–202, https://doi.org/10.1016/S0196-1152(07)15005-5.

9. Turner and Gardner, *Landscape Ecology in Theory and Practice.*

10. Kelley A. Crews-Meyer, "Temporal Extensions of Landscape Ecology Theory and Practice: Examples from the Peruvian Amazon," *The Professional Geographer* 58, no. 4 (November 1, 2006): 421–435, https://doi.org/10.1111/j.1467-9272.2006.00579.x.

11. John R. Stilgoe, *What Is Landscape?*, Reprint edition (Cambridge, MA: The MIT Press, 2018).

12. Thomas Dunne and Luna B. Leopold, *Water in Environmental Planning* (New York, NY: Macmillan, 1978).

13. Hyoungsu Park and Daniel T. Cox, "Probabilistic Assessment of Near-Field Tsunami Hazards: Inundation Depth, Velocity, Momentum Flux, Arrival Time, and Duration Applied to Seaside, Oregon," *Coastal Engineering* 117 (November 1, 2016): 79–96, https://doi.org/10.1016/j.coastaleng.2016.07.011.

14. Henry Hooper, "North Oregon Coast: Clatsop Indians," *Henry E. Hooper* (blog), August 7, 2013, https://henryehooper.blog/north-oregon-coast-clatsop-indians/.

15. Douglas Deur, "The Making of Seaside's 'Indian Place': Contested and Enduring Native Spaces on the Nineteenth Century Oregon Coast," *Oregon Historical Quarterly* 117, no. 4 (2016): 536–73, https://doi.org/10.5403/oregonhistq.117.4.0536.

16. Ian L. McHarg, *Design with Nature*, 25th edition (New York, NY: Wiley, 1995).

17. Stefan Leyk, Johannes H. Uhl, Dylan S. Connor, Anna E. Braswell, Nathan Mietkiewicz, Jennifer K. Balch and Myron Gutmann, "Two Centuries of Settlement and Urban Development in the United States," *Science Advances*, June 2020, https://doi.org/10.1126/sciadv.aba2937.

18. Angie Venturato, *A Digital Elevation Model for Seaside, Oregon: Procedures, Data Sources, and Analyses* (Seattle, WA: NOAA, 2005), https://repository.library.noaa.gov/view/noaa/11061.

CHAPTER 2

Landscape information: the problem of analysis

All the information was there: 300 different opinions on what the future of Ravensdale Park should be. At the community meeting, families with young children put stickers on our boards to indicate their preference for various landscape elements. Children in the upper portion of the gym played soccer. Other kids just outside tossed a baseball. The data overwhelmingly pointed to the community wanting athletic fields for their children.

I had completed a site analysis. At the meeting, we displayed maps showing the regional sports fields and the gap in fields that existed in this particular area. Community members strolled past brief histories of the town and its coal mining past. I presented a series of potential "spaces" that might go in the future park (working closely with the County) that responded to perceived community needs. I knew the distance of buffers required to protect the streams on adjacent property. I left the meeting happy, excited about the tidal wave of support for the project and ready to start designing and … I had completely missed the mark.

The first suspicion something might be wrong arose during the very next meeting. It was not a public meeting, in the sense of a notice from the County opening up the discussion of Ravensdale Park to anyone and everyone. It was a small meeting of a few different groups: environmentalists, soccer interests, Little League, local officials and the school district. Given the mandate of the last community gathering, I brought a series of slides and discussion points to hash out neighborhood issues like field lighting, field use, opening and closing times, and parking. When I walked in, there were 12 people around the conference table I did not recognize. As we sat down, one of them yelled something about the corrupt power of the County. And we were off. A lot was said in that

DOI: 10.4324/9781003158943-3

2.1 Ravensdale site analysis collage, a muddle of information.

meeting … some of it spiteful, some kind, but it was clear that a small group of residents who lived adjacent to the park had not been heard. They lived in houses built by the coal mining company one hundred years ago or in the trailer park across the street or in large lot homes seeking refuge from the city. They did not want sports fields. They saw this space as a community gathering area (though they did not yet have an idea of what that might look like). My initial lack of understanding of how the history of the place informed contemporary living and place prevented me from seeing the dynamic nature of the landscape.

We think we know. We gather information – from the basket of nature and the reservoir of culture – and we assemble this information into a picture of the site's opportunities and constraints. As humans, information gathering is what we do. Information, whether collected from maps, histories or other people, permeates the experience of place, making it a crucial part of landscape dialogue. However, information gathering alone may miss the spatial and experiential that is difficult to capture in an analytical process. Landscape dialogue attempts to flesh out these experiences over time to understand the complexities of a place. Where information gathering collects *spatial* data for GIS, dialogue examines *space* as a lived and relational encounter. What does this mean? It is a conversation with space, an often unconscious assessment of the quality of the blank, but not blankness, between the two trees, along the sidewalk or within the vacant lot. This space can be sacrosanct; although it looks blank to the visitor, it is filled with meaning from the past, with conflict and with hope for the future.

After the contentious Ravensdale Park meeting, the County parks planner and I arranged a series of meetings with a few community members in a hair salon next to the trailer park. This was not about gathering information, although there was some of that. It was about dialogue. It was about listening, a lot of listening … to concerns, histories, struggles and anger over the "big town" of Maple Valley not building their own parks. It was about speaking a more inclusive

2.2 Proposed community open space at Ravensdale Park.

vision. We continued to meet with the baseball and soccer advocates as well. Both groups were calling their Council representatives with varying degrees of hysteria so there was real pressure. In the end, the baseball person (who was the most local of the sports advocates) called the environmentalist and a local resident and hashed out a compromise that divided the park into two: sports fields on one side and community open space on the other. It was never about information – on the history, on the needs, on the ecology of the park; it was about people and their inhabitance of the space.

Given the challenges of the information/rational approach to site analysis, how should designers incorporate information into the process of landscape dialogue? In this chapter, we delve into the role of information as a multi-dimensional aspect of the landscape (with strings attached).

LANDSCAPE IS INFORMATION

In a logical sense, dialogue is the exchange of information. There may be times when dialogue is *only* information. When I call my spouse on the way home and ask what we need to get at the grocery store, I am asking for a list of items, not inviting a discussion of the quality of our relationship. That can wait (usually!) until I get home and we can dialogue face to face.

In academic circles, information (or data) is divided into two categories: quantitative and qualitative. Quantitative data encompasses that which can be counted or calculated: the percentage of visitors who identify as Asian American, the area of undisturbed riparian forest or a projected curve of supply and demand. In contrast, qualitative data is a measure of the quality or character of something, the experience of a skateboarder in a plaza, the stories told about a neighborhood's history or the preferences of an elderly person's transportation mode. Often opposed, quantitative and qualitative information overlap, i.e. the visual preference surveys introduced in the next chapter where an individual's preferences are aggregated with others to provide a picture of what the general population prefers. Both quantitative and qualitative processes organize data into categories, sorted, classified, evaluated and presented. Designers do this constantly, categorizing information into useful or not useful, good or bad, time-consuming or efficient, personal or collective. And the best designers use this information to create great designs.

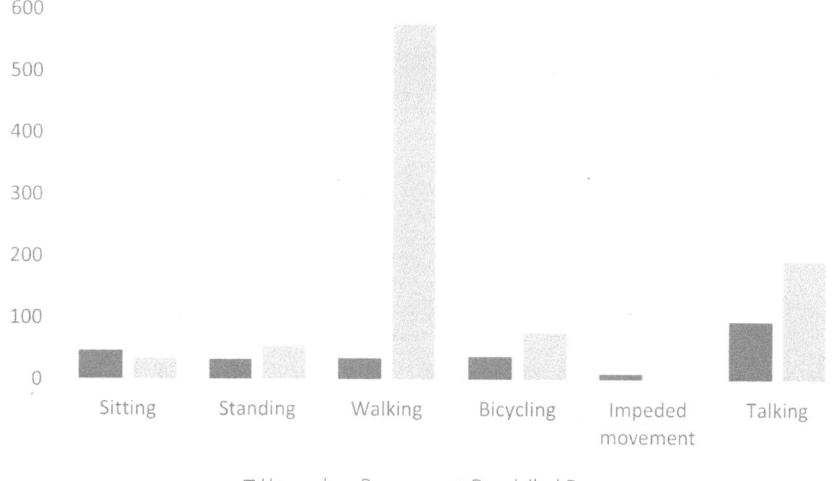

2.3 A graph of quantitative information – recording behavior of people along Pacific Avenue in Santa Cruz, California. The majority of people appeared to be domiciled and walked along the sidewalk to shop. More unhoused people sit (on benches or on the curb) than domiciled people.[1]

Economists' rational choice theory attempts to explain how people use information to make decisions.[2] The theory posits that cities or landscapes (or really anything) are formed through an aggregate of individual choices. Individuals make choices by first assembling *all* available information, then processing it to maximize benefits and minimize costs in a social cost–benefit analysis, e.g. applied to economic behavior.[3] In transportation, for instance, if an origin offers fewer benefits than a destination, movement occurs.[4] Information, in this model, provides decision-makers a suite of options, each rated as a cost or benefit. The more information that is processed, the better the decision.

The landscape contains a plethora of information: subsurface composition of rock and soil, the habitat of songbirds, the movement of children as they gather for play, the time of increased traffic flow, the opinion of people walking in the dark. We divide landscape information into categories to learn about a place (Table 2.1). Each category relates to a discipline with a body of knowledge and particular ways of working.

The well-established McHargian process of layering information, assigning value then determining potential use, can be seen as a kind of spatial rational planning.[5] Planners assemble information about a site as layers, then evaluate each layer for its cost or benefit in relation to development. By treating the landscape as an "economy," a weighing of costs and benefits, the analysis and future re-configuration of that landscape is rational, a "natural" way of assessing a place.

Table 2.1 Typical categories of landscape information.

	Discipline	Layer
Natural conditions		
	geology	topography
		aerial photographs
	soil science	soil depth, type
		soil drainage
		soil erosion
	hydrology	climate and rainfall
		groundwater
		surface flow
		watersheds
		flood mapping
	ecology	plant communities
		plant species
Cultural conditions		
	geography	development patterns
		density
	planning	land use
		zoning
	economics	income levels
	sociology	race and ethnicity
		crime and policing
		age and family groups
	political science	party affiliation
	history	original people groups
		stories of development

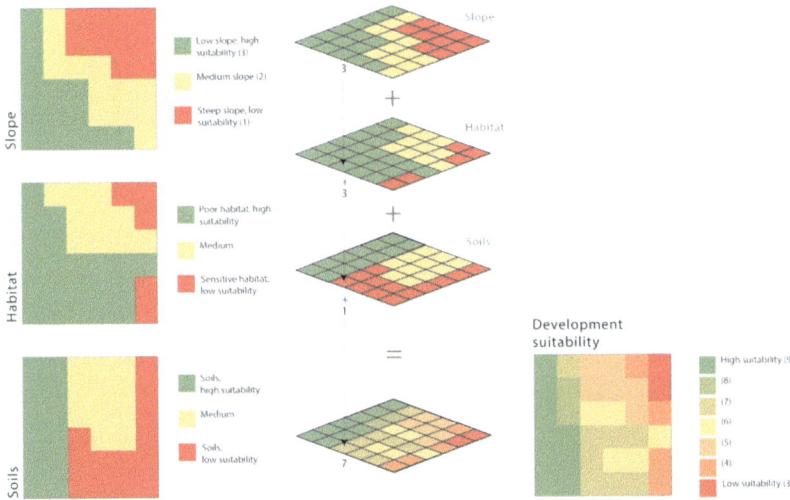

2.4 McHargian layers of slope, habitat and soils with ranked suitability mapped, then added together to generate a suitability map of development.

LANDSCAPE IS NOT INFORMATION

Yet humans are not rational. The landscape is not an economy. Portions of land may be bought and sold but reducing landscape to property or site misses its deeper structures and processes that shape and are shaped by our everyday lives. The idea that we can assemble all available information about a site, analyze it, and then use that analysis to inform the best design does not reflect the realities of landscape or design. Information analysis does not account for: (1) the hidden assumptions we have about the landscape; (2) the integrative nature of landscape; and (3) a certain blindness arising from the need for control. I take each of these in turn.

First, it is not individual, rational choices that construct and maintain the landscape but our collective agreement with assumed ways of organizing space. Pierre Bourdieu, a French sociologist, rejected rational choice theory as a model for how people actually make decisions, pointing instead to the primary role of social ties and norms that guide us in everyday living.[6] Classifying the world according to social norms removes the need to constantly assess our environment, an impossible task given the amount of information available, so we may rely on existing assumptions of "this is how the world works." As we assess a landscape for future design, we must confront these social assumptions if we are to understand a place. Bourdieu studied small Berber villages in Algeria, North Africa. His "analysis" of the villages has a spatial component, an unquestioned arrangement of buildings that reinforced existing village hierarchies of chief and peasant, man and woman. To assess this important social landscape first requires an understanding of these hierarchies, an examination of assumptions of both village resident and anthropologist.

In contrast, the analytical approach reduces an integrated whole to a series of more easily understandable parts, which may exclude social hierarchies. This works well in understanding specific information related to already dis-integrated disciplines. It works less well when the majority of people's behavior stems from social, normative agreements.

Second, landscape works as integrated flows, not static layers. Site analysis examines separate layers of information. When it comes time to assemble the information back into a semblance of the original landscape, designers rely on synthesis – the process of reassembling. It turns out to be difficult to rejoin the localized ecologies, economies and cultural signifiers so as to make comprehensive statements about a place. At times, it may

Town gateway
(men)

Main square
(family)

Well
(women)

2.5 Spatial analysis of traditional hill town in Algeria.

be necessary to look at a single landscape element (always in the context of a larger whole), but if it is possible to assess a place, its meanings and forms as a whole, designers can increase the efficacy of assessment. It takes the more holistic sciences, like ecology, and the eyes (and ears) of an artist to do so.

Information separates. It does not integrate. The perilous distinctions shape the landscape as we divide nature from culture and discipline from discipline. We divide, not only the information, but the ways of knowing. And the result, as Berry states in "Poem VI," is losing an "order we are ignorant of":

The intellect so ravenous to know
And in its knowing hold the very light,
Disclosing what is so and what not so,

Must finally know the dark...[7]

2.6 Forest collage, moving through it, reading it and surveying it.

Third, the designer's motivation emerges as a nebulous but important factor in landscape relations. If the designer desires to impose a grand vision onto a place, onto its surface, then the first task will be to extract information from it. If the motivation is how a design will look on a magazine cover or Instagram shot, then there is little need to understand a place, certainly to the extent that the design responds to the stories and processes within the landscape.

Mulholland considers this type of "reading" a form of control.[8] We select the text. We know what we want to get out of it. "We read the text analytically, viewing it as an object over which we as subject exercise our control, to ensure that it conforms more or less comfortably to our purposes."[9] We are information readers of the landscape. What makes the analytical process subject to control? When we select part of the landscape – the vegetation, say – we place that information in a discipline; we "discipline" the information. And whether we treat it as a list of plant species, a plant community or divide it into additional categories (i.e. invasive and native), we exert the discipline's prescribed methods of analysis to make statements about the vegetation that stay within established parameters. The information does not challenge or inspire us towards other purposes. We are left as the same designers we were when we began (see Chapter 8).

If landscape analysis becomes landscape dialogue, we awake to the peril of an accumulation of information which makes it unnecessary to listen to the landscape itself. In conversation, we stop listening when we assume we know what the other person is going to say. Our expectations about a landscape frame how we see it. We "project," in the classic psychological sense of the term, to fill the empty gaps in our knowledge with our own ideas, as people do when observing a painting or sculpture.[10] While landscape information is necessary to understand a place, the analytic processing of that information into categories, the reduction to simplistic statements, misses the opportunity for dialogue and understanding.

LANDSCAPE INFORMATION AS LISTENING TO THE LANDSCAPE

What then should we do with information? How do we wrestle with landscape information, in the form of maps, stories and sensory experiences, holistically, without disassembling to the point of misunderstanding? In landscape dialogue, we still gather information, but with the understanding it is not like one gathers eggs from a chicken coop and more like one absorbs a fried egg at breakfast. Information exists as an integral part of the landscape and our experience and memory can be employed to remember it, to listen to it, to engage with it.

Swimming in information, we do not completely discard analysis but recognize it is more than description. As anyone who has taught site analysis knows, it is very difficult to inspire students to move beyond description to analyze an aspect of the landscape. Analysis draws from the information and makes a statement about it. It does not regurgitate that information. A clay/loam soil filled with construction debris becomes a soil unsuitable for building. An area of the site that sees bicyclists congregate in the morning becomes an important trailhead connected to a larger network because of its position at the edge of the city. Analysis makes an evaluative statement related to future design. Preserve this, remodel this, remove this other thing. Dialogue forces one to evaluate as one listens and speaks.

Early in my career, during the U.S. 93 highway project in Montana, I set up a series of site meetings with various stakeholders and the design teams at the location of each potential wildlife crossing, often low-lying areas we had already identified as places where wildlife crosses the existing road. The highway engineers wanted to reduce or eliminate each crossing to save on project costs, prioritizing human travel by car, while we, on the same design team, wanted to add crossings in support of a more porous highway (see Chapter 5). A team of landscape architects from our office led the effort in the field. I had outlined strategic information we wanted to know at each location, like the size and material of existing culverts, plant species, etc. I had printed out forms for each place with blank spaces for measurements. We were ready. When we arrived in the field, the design team and client representatives began talking. It was quickly apparent that no one from our office would be collecting information. They wanted to debate the size of the crossings with the engineers and wildlife biologist. The work of examining the landscape systematically was neglected (or I did it myself). It was an acknowledgement that important decisions are made (the size and location of wildlife crossings) talking to others, not gathering data. The data would help (I maintain) in the long run, but the hard work of measuring and evaluating did not compensate for missing out on weighing in on the dialogue.

Designers are generalists. Ideally, we learn about each discipline with special attention to a few like ecology and sociology, so we can make connections during the act of perception. *Synthesis occurs, not as a future layering, but as part of perception through a gathering of disciplinary experience.* Here, Wendell Berry would make the point that this "gathering of experience" takes a lifetime of living in one place, tilling the same landscape. An excerpt from the poem, "In a Motel Parking Lot, Thinking of Dr. Williams," makes the point:

> For want of song and stories
> they have dug away the soil,
> paved over what is left,
>
> set up their perfunctory walls
> in tribute to no god,
> for the love of no man or woman
>
> so that the good that was here
> cannot be called back
> except by long waiting.[11]

A deep understanding of place and what might be not only possible, but beneficial, requires a "long waiting." This leaves the designer in a bind, attending to the limits and possibilities of time. The architect must consider it possible to shorten the time frame of understanding through other means – experiences of landscape elsewhere, a continuous study of methods and techniques – otherwise, healthy design is impossible except in the context of a life-long rootedness in the landscape.

In V.S. Naipaul's *Enigma of Arrival*, he describes his entrance into the Cornish landscape of England after growing up in Trinidad. He writes of a gradual awakening to the place as he takes daily walks from his new home. Eventually, the "forest" that forms a backdrop of his walks reveals itself to be only a small grove of trees between his house and the neighboring farm. The wide open, unoccupied spaces of field and lane become a cultural palimpsest of generations of farmers and families residing in this place. He focuses in on the changing nature of color and growth and fertility of his neighbor Jack's garden.

I saw with the eyes of pleasure. But knowledge came slowly to me. It was not like the almost instinctive knowledge that had come to me as a child of the plants and flowers of Trinidad; it was like learning a second language. If I knew then what I know now I would be able to reconstruct the seasons of Jack's garden or gardens. But I can remember only simple things like the bulbs of spring.[12]

He learns, as an observant writer, that the lane he walks is called a "droveway," following an ancient river valley. A droveway was a larger path, 12 meters wide, for driving cattle to markets; a practice that had stopped by the 1800s with the arrival of trains. He revels in the openness of the place but, during the first year, someone places a barbed wire fence along one side. And he perceives "that I was also in a way at the end of the thing I had come upon."[13] Time spent walking allowed him to move from the descriptive to the analytical to a dialogue. Past and present, openness and enclosure, each landscape process shapes his outsider viewpoint to be more like the landscape itself as a changing and dynamic entity.

Landscape information becomes integrated into the process of landscape dialogue through time, an openness to experience and an application of accumulated information on the experience itself. The process moves circularly between absorbing information of a place (often away from the place) and applying that information in the field. That a landscape is dominated by red alder, an early successional species, for instance, leads us to examine the soils and ground for scars of disturbance and then go back to aerial photographs to uncover the source and timing of that disturbance. That a plaza has an imposed curfew of 8 pm inspires us to observe the park at 7:45 pm and then review newspapers to find civic concerns related to homelessness, panhandling and drug use that might account for the fear of night occupation. Information as spark, as framework for seeing.

 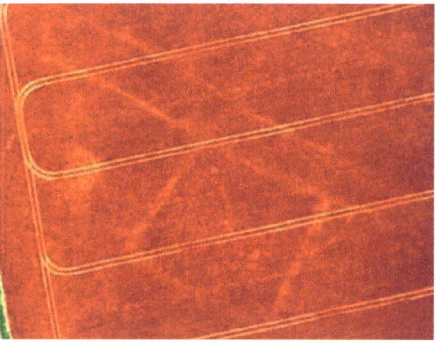

2.7 Aerial archaeology uses drone photography to show linear patterns of ancient droveways through agricultural fields in Yorkshire, England. Archaeologists use false color to highlight the ancient infrastructure, which lies beneath contemporary tractor marks.

THE MANAGEMENT OF PARKS

The U.S. National Park Service (NPS) has developed one of the most robust processes for gathering information on the landscape that currently exists. One hundred years of experience evaluating the historic, natural and recreational "layers" of the scenic parks informs the process. They have made many errors that gradually have informed the public process, making it better and more comprehensive. The NPS process illustrates the advantages and disadvantages of a rational, informational approach.

The original Organic Act of 1916 that established the NPS in the United States centers and structures the "management" of the parks – the dual mandate of conserving "the scenery and the natural and historic objects and the wildlife therein" while providing "for the enjoyment of the same in such manner and by such means as will leave them unimpaired for the enjoyment of future generations" (Sec.1).[14] Already in 1916, the Organic Act framed the dialogue between people and landscape as one of management arising from a sense of stewardship, or even responsibility, to be found in NPS staff. A dialogue between management and employees is hierarchical, a power dynamic that filters the exchange of information (often in unacknowledged ways). The Park Service developed a Management Policy document over the years that guides almost everything they do – a guide to management of the parks.[15] It covers the protection of natural and cultural resources, public planning, facilities and use of the parks. On page 24, the public engagement process is outlined:

Public involvement will meet National Environmental Protection Act (NEPA) and other federal requirements for

- identifying the scope of issues,
- developing the range of alternatives considered in planning,
- reviewing the analysis of potential *impacts*, and
- *disclosing the rationale* for decisions about the park's future.

The Park Service will use the public involvement process to

- *share information* about legal and policy mandates, the planning process, issues, and proposed management directions,
- learn about the values placed by other people and groups on the same resources and visitor experiences, and
- build support for implementing the plan among local interests, visitors, Congress, and others at the regional and national levels.[16]

The Management Policy document can be considered a dialogue between the agency and the public, between the agency and the parks. As such, it is significant that the word "landscape" does not appear in the description of process. Several key phrases situate their approach within rational choice theory. Human interaction with the park is as a visitor who "impacts" the landscape. The public will be won over with an explanation of clear alternatives, an analysis and then "disclosing" (at the end?) the rationality of the (Park Service's) choice based on information. The public process is not described as a dialogue, but as something to be "used" to "share information" from the Park Service, to learn about values, and to build support for the plan – a process that might include dialogue, but dialogue limited to information exchange. Even when using the language of values, the Park Service must remind us that "resources" are limited and not everyone is going to get what they want. As described, public involvement is a distinct process, one step in the larger management of the parks, separating "the public" from the landscape (and also from NPS staff who operate, not as part of the public themselves, but as caretakers for the public).

The goal of NPS park management outlines an attempt to be equitable, to involve the public in a transparent process and provide clear reasons for decisions made. Admirable goals. However, the text itself provides a rationale for management as the primary goal.[17] Because dialogue in this case equates with information exchange, the NPS public process exists to facilitate this information exchange in ways that perpetuate the agency and the process; the Organic Act becomes secondary. The linear process, the documentation, become ends in themselves – tools of separation from the visitors' experience and their emotions. A list of public meeting attendees is of the utmost importance in this process to show who and how many people "weighed in" on park decisions. It gets filed in an office, so that when the process is legally challenged, the page can be produced and the information-exchange process shown to be complete. A document with a list of attendees as information has little relation to the landscape of the parks. In my experience, one of the unstated goals of the process appears to be to *avoid* listening to people, in particular their emotions about a place. Taking attendance paradoxically justifies dismissing attendees. The separation of Park staff who make decisions from the public and also from the rangers working in the landscape leads to a separation of the process from the park landscape and its context. Rational choices can then be made.

The NEPA Process

2.8 A decision-making flow chart for the National Environmental Protection Act (NEPA) assessment used by the National Park Service. The "appropriate" public process steps have been highlighted (which brings up the question: what is an *inappropriate* public involvement in *public* lands?).

Landscape dialogue exchanges information but, like a conversation of some depth, uses that information to inquire and understand. Assessment then includes shared experiences.

PRAXIS

In landscape dialogue, the challenge is to gather information (listening) while holding it as part of a whole to then communicate findings (speaking) back to the landscape community to see if one understands. To absorb landscape information in this way embraces measurement and other methods of traditional site analysis, with the understanding that measuring is a process of understanding and not a data point (or not just a data point). Books, such as LaGro's *Site Analysis*[18] and Marsh's *Landscape Planning*,[19] offer useful compendiums of site measurement in the procurement of landscape information. The following practices emphasize the measurement of space *as experienced*.

1. Measuring space with the body

As one practices spatial measurements, the body becomes the most sensitive instrument of assessment. Measure your pace. Walk a line with a pre-determined distance and count your steps. Divide the distance by the number of steps to determine the length of your pace. Now, measure the "site," walking from one boundary to the other counting the number of steps to yield a distance. Of course, this distance can be more accurately measured on a map or site plan but this basic measure gives the designer a more immersive way to engage with spatial information as one walks through the landscape, its ups and downs, its lines and curves. This will help the designer with elements of scale in the future design. For larger landscapes, walk the site from end to end and note the time it takes. Then, calculate the distance of the walk based on the average speed of your walk, accounting for terrain.

Use your pace to measure *spaces* within the landscape from edge to edge, i.e. building to building, tree trunk to edge of tree canopy. The selection of which spaces to measure becomes part of the process. Some spaces may extend off site. In plan view, sketch the difference between the site itself and the spaces within and beyond it.

Assess the conditions of the edges of space. Are they hard, like a building wall, or porous, like a row of bollards or trees? Diagram the spaces you measure; show how they overlap or are contained within the site.

2. Measuring diversity

Diversity is originally an ecological concept, a measure of difference over a defined space. For example, a monoculture of one plant, as in agriculture, contrasts with a complex ecosystem of biological richness, such as a temperate rainforest. In all but the smallest defined spaces, it would be overwhelming to measure all elements, i.e. recording all plants within a forested landscape. Therefore, ecologists rely on methods of randomization in the field to summarize or "sample" the terrain. The challenge is to choose the sample carefully so that it represents the true plant community within the area. Examples of sampling a plant community include the point-centered quarter method in which the area around a random point is divided into four quarters of the compass and the distance to the nearest tree or shrub's base is measured, along with the basal area.[20] Total density of plants can be estimated from these measurements.

Other aspects of the landscape can be measured for diversity. Hydrology is a critical component of landscape, a "layer" that consistently stands for the health of a larger system. Use the pebble count method to measure the size and diversity of the sediment on the bottom of a stream or river.[21] Step into the water; pick up a particle/pebble/stone at the point of your toe; measure the size of the particle in millimeters; record the measurement. Then step elsewhere and repeat. Do this 100 times (or 50 and extrapolate). Geomorphologists do this to systematically describe the composition of the stream bed. It has the added benefit of immersing one's body (or at least your feet) in the water. The sediment composition speaks of a landscape's topography, the power of the stream or river and the cycle/frequency of flood events.[22]

Diversity can be applied to human behavior as well (although somewhat more problematically). Measurements of cultural or racial diversity in public space often misrepresent how people would describe themselves, relying on the assignation of racial categories based on appearance. An analysis of cultural diversity in public space should use ethnographic methods which utilize time and relationships in the landscape to evaluate space.[23] However, behavioral diversity can be measured without long-term time commitments. Count people in a public space and record their different behaviors. Select a plaza or open space and set boundaries for the study of behavior, i.e. from street to street. Use a pre-drawn map and clipboard. Begin facing one direction and record on the map

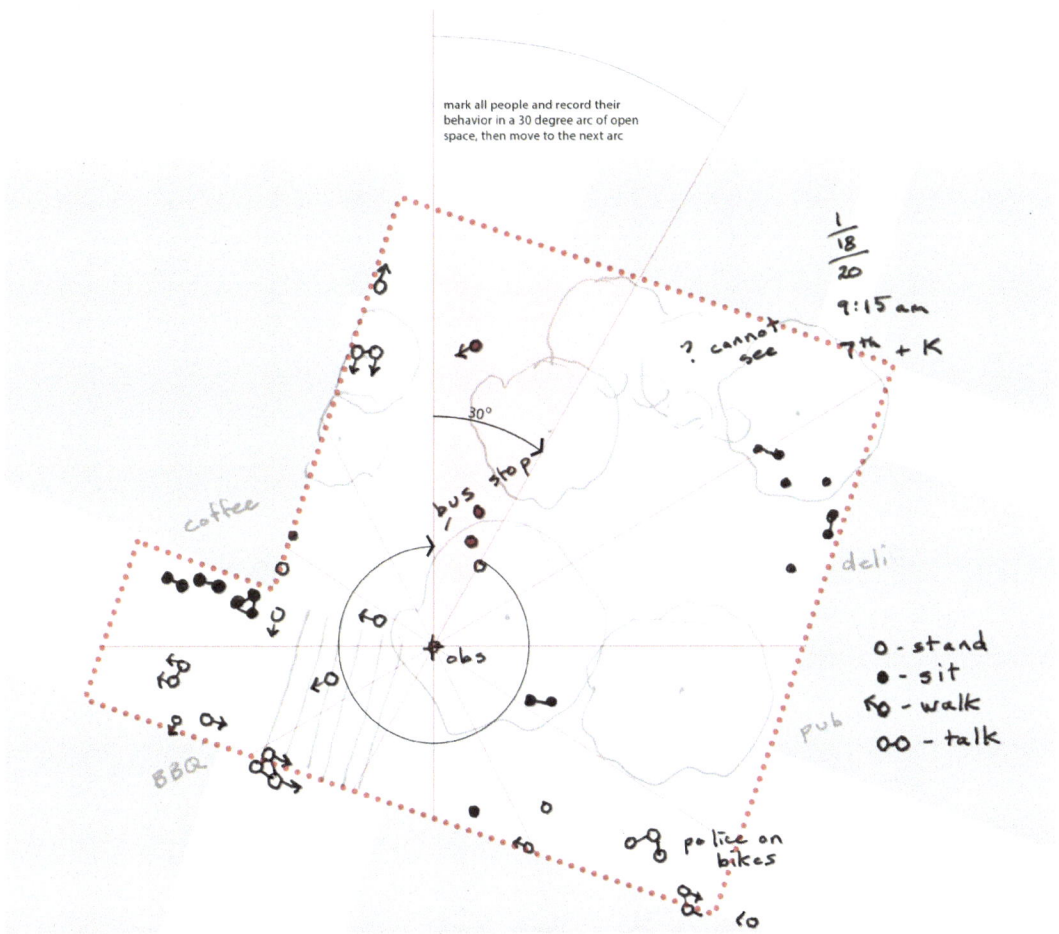

mark all people and record their
behavior in a 30 degree arc of open
space, then move to the next arc

2.9 Behavior map of urban plaza. Base map showing buildings printed beforehand, then recording people and their behaviors from a fixed point clockwise through the public space.

every person's location and their behavior in a swath that extends from the observer to the established boundary. Behavior can include sitting, standing, walking, holding hands, talking on the phone, etc. It is helpful to develop a symbology with dots, circles, arrows, etc. to be able to write down behaviors quickly. Then turn 30 degrees clockwise and record the behaviors one sees in this new direction. Continue observing and recording swaths of the landscape clockwise for 360 degrees. If an observed person walks from one observational swath to another, record them twice. (Using the same logic, you will miss some people who will be walking counter-clockwise.) When finished with your observations, the behavior map will show a sampled map of behavioral diversity that can be used to inform design.

3. Translating measurement in the office/studio

After measuring space in the field, how can these measurements inform the design of a future landscape with its potential complexity of soils and pavement and plants and people? And, when returning to the office, how does one hold the experience of the place as central to a continuing dialogue?

Transition time to connect field to office

The office/studio will be the place where we assemble the sketches, diagrams, aerial photos and photographs. It will serve as a repository of landscape information. It is important to keep the information together, to keep the relationships at the forefront (i.e. between soil and plants, people and pavement). To this end, set aside a transition time moving from field to office. If the office is close to the field, commandeer a conference room or big table away from others. If the office is far from the field site, locate a nearby coffee shop. The goal is to write down thoughts from the field as soon as possible after your experience. These thoughts do not require examination at this point, just a stream of consciousness record of what stood out about the place. Use this transition time to hold together different site elements. Ask "why?" questions of the place that can be answered later: i.e. why were people passing through the plaza and not stopping or why has this site avoided development in the past? The questions will be a useful guide to the next phase of evaluation.

Assembling a physical base

The studio does have advantages: access to design materials, digital information on wide monitors, other designers who can give feedback, flat surfaces to spread materials out. Once back, assemble the important information in a physical base. As in conventional site analysis, this often begins with an aerial photo, an excellent organizer of site information as it contains so much information regarding space, materials and measurements. (It may not address immersive space and changes over time.) Print out the largest aerial photo of your site possible. Show an area larger than the property/site. Then layer the sketches, digital information on soils and plants, and the bodily measurements you have made on top of the aerial forming a hybrid base of information related to your plan. Then, write notes addressing landscape questions and the arrival and leaving of the site onto the plan or sheet. Use trace paper. This works best as a group activity.

How is this different than a McHargian layering common to GIS analysis? The physical base could be done in GIS, but there is an aspect of sketching, of using one's hand and layers of trace paper, of shaping a collage that rejects hard boundaries (polygons) and injects qualitative "data" into the dialogue. The combination of sketches and digital information will be a visual illustration of the dialogue up to this point of design. Create this base or model as a tool for engaging with the landscape during the duration of the project, so that it is updated as new information emerges.

Which information is important? Use a landscape's spaces to organize the information. Sketches of space. Histories of space. After plan view analysis, switch to sections. Draw the edges of the different spaces in section. Ignore legal boundaries. Stay inaccurate. Cross property boundaries. Overlap polygons. Use gestural sketching. Choose charcoal or another rough media if necessary. Force yourself to transgress established boundaries within the measurements and recordings themselves. In later design phases, there will be time for accurate assessments of where elements may go.

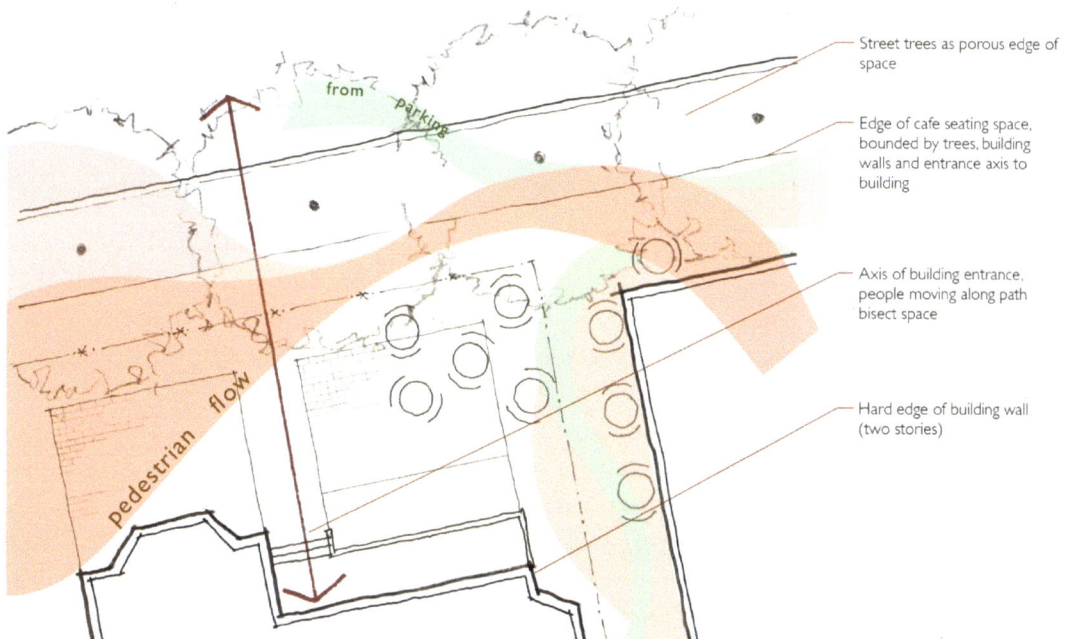

2.10 Landscape sketch/diagram of small urban space.

NOTES

1. See Cory Parker, "Homeless Negotiations of Public Space in Two California Cities" (University of California, Davis, 2019), https://escholarship.org/uc/item/9x77627p. Distinguishing between people experiencing homelessness and domiciled people is problematic. I discuss parameters for estimating different populations in the thesis.
2. André Blais, *To Vote Or Not to Vote? The Merits and Limits of Rational Choice Theory* (Pittsburgh, PA: University of Pittsburgh Press, 2000).
3. Blais, *To Vote Or Not to Vote?*
4. John Carl Lowe and S. Moryadas, *The Geography of Movement* (Prospect Hights, IL: Waveland Press Inc, 1984).
5. Ian L. McHarg, *Design with Nature*, 25th edition (New York, NY: Wiley, 1995).
6. Pierre Bourdieu, *Distinction: A Social Critique of the Judgement of Taste* (Cambridge, MA: Harvard University Press, 1984).
7. Wendell Berry, "Poem VI," *A Timbered Choir: The Sabbath Poems 1979–1997*, 1st edition (Washington, DC: Counterpoint, 1999).
8. Robert Mulholland, *Shaped by the Word: The Power of Scripture in Spiritual Formation* (Nashville, TN: Upper Room, 1985).
9. Mulholland, *Shaped by the Word*, 94/171 Adobe Digital Editions.
10. E.H. Gombrich, *Art and Illusion: A Study in the Psychology of Pictorial Representation* (Oxford: Pantheon, 1961).
11. Wendell Berry, "In a Motel Parking Lot, Thinking of Dr. Williams," in *New Collected Poems* pp. 317-318. (Berkeley, CA: Counterpoint, 2013), 317–318.
12. V.S. Naipaul, *The Enigma of Arrival*, Reprint edition (London: Vintage, 2010), 33.
13. Naipaul, *The Enigma of Arrival*, 26.
14. United States Congress, "National Park Service Organic Act," 16 U.S.C. 1,2,3 and 4 § (1916).
15. National Park Service, *Management Policies 2006* (Washington, DC: Department of the Interior, 2006), www.nps.gov/policy/MP_2006.pdf.
16. National Park Service, *Management Policies 2006*, 24; italics added.
17. To the credit of the National Park Service, many of their staff, in particular park rangers, are incredibly dedicated to the dual goals of the Organic Act. For a review of NPS attempts to assess their own agency and the Park System, see Jane McDonnel's "Reassessing the National Park Service and the National Park System," *The George Wright Forum* 25, no. 2 (2005). For an entertaining account of NPS mismanagement, see Alton Chase's *Playing God in Yellowstone: The Destruction of America's First National Park* (San Diego, CA: Harper Paperbacks, 1987) and a more contemporary account from a former park ranger: Jordan Smith's *Engineering Eden: A Violent Death, a Federal Trial, and the Struggle to Restore Nature in Our National Parks*

(New York, NY: The Experiment, 2019). The process of an agency established for a specific purpose eventually assuming the very practical goal of perpetuating the agency itself, rather than its original purpose, is in no way unique to the Park Service.

18. James A. LaGro, *Site Analysis: Informing Context-Sensitive and Sustainable Site Planning and Design*, 3rd edition (Hoboken, NJ: Wiley, 2013), http://site.ebrary.com/lib/ucdavis/Doc?id=10653568.

19. W.M. Marsh, *Landscape Planning: Environmental Applications*, 4th edition (Hoboken, NJ: Wiley, 2005).

20. See Chapter 9 of Michael Barbour, Jack Burk, and Wanna Pitts, *Terrestrial Plant Ecology*, 2nd edition (Menlo Park, CA: Benjamin/Cummings Publishing, 1980).

21. Gregory S. Bevenger, *A Pebble Count Procedure for Assessing Watershed Cumulative Effects* (Washington, DC: U.S. Department of Agriculture, Forest Service, Rocky Mountain Forest and Range Experiment Station, 1995).

22. See Kirstie A. Fryirs and Gary J. Brierley, *Geomorphic Analysis of River Systems: An Approach to Reading the Landscape*, 1st edition (Chichester: Wiley-Blackwell, 2012). The classic text on hydrology remains Thomas Dunne and Luna B. Leopold, *Water in Environmental Planning* (New York, NY: Macmillan, 1978).

23. For example, Setha Low, *On the Plaza: The Politics of Public Space and Culture*, 1st edition (Austin, TX: University of Texas Press, 2000). See Low's rapid ethnography method as well.

CHAPTER 3

Landscape perception and motion

> Landscape perception is not an analysis of a picture, but an experience of motion.

Landscape perception – the sensory experience of a scene – has been synonymous with *visual* analysis in the mind of the designer. For many years, research in landscape perception was dominated by visual preference surveys. The work of Rachel and Stephen Kaplan, for instance, summarized in the book *With People in Mind*, offered perceptive insight into human–environment relations, such as the preference for water in a scene or a curving path disappearing into the distance.[1] Visual preference studies attempt to quantitatively answer questions regarding people's opinions on scenery by showing them slides and asking them to rate the scene on a Likert scale from 1 to 5, 1 being "do not like" to 5 being "love it!" With enough participants, one can make generalizations about what kind of scenery people prefer. This added rigor to the fuzzy field of aesthetics. Government agencies appreciated the rigor, including visual preference surveys in large-scale environmental impact statements.

For many years, the study of landscape preference revolved around the idea of scenery. Why is this landscape considered scenic? What makes it picturesque? And how do different people, say visitors, farmers, laborers, respond to this landscape in different ways? At the root of these questions lies human perception – how people sense the world around them. Landscape perception is not a static, visual activity. We do not navigate our world through mentally processing snapshots taken of scenes. We negotiate the world as sensing people immersed in the world. And negotiation it is. A constant give and take between a human moving within their environment. And this shapes perception. Take the example

DOI: 10.4324/9781003158943-4

a)   2.64   b)   3.76   c)   2.15

d)   3.58   e)   2.88   f)   3.67

3.1 Six sample photographs (scenes) graded for attraction on a Likert scale. Higher scores were given to photographs with a curving path (b), open space with clear understory (d), and scenes with water (f), after Kaplan and Kaplan's findings.

of a red barn, a culturally significant object in the American landscape. People enjoy looking at a red barn.[2] The traditional red barn is a static relic, without a functional purpose in contemporary agriculture. It has been replaced by the steel framed warehouses of dairy cows and machines, as well as the infrastructure needed for direct-to-market produce. Visual preferences alone are not going to explain the appeal of the red barn. The pastoral landscape of row crops, fence and hedgerow, a few scattered sheep in a pen and a red barn do not just present a pleasing mix of color and light to our eyes but evoke memories of what a landscape used to look like. "As a symbol, the red barn icon represents common human values, such as the nostalgia of a simpler era, and the benefits inherited from the past."[3]

While visual preference surveys may tell us important things about how people view a picture or a photograph, they tell us considerably less about how people experience a landscape. Landscape perception is a larger and more complex process than viewing an image. In the study of perception, psychologists, evolutionary biologists, architects, ecologists and philosophers examine the way humans sense the world. They debate the primacy of vision, the role of emotions and the right balance of light

3.2 Scenic view of a red barn in the Central Valley of California.

and dark, tone and shading. Designers can look to this work for insight into landscape dialogue.

To understand the process of perceiving the landscape through dialogue, I use the red barn as a fulcrum for the visual, the experiential and the assignment of value to a landscape. How do we see the red barn in the landscape? What is our experience of it? And what does it symbolize? Explanations of landscape perception should address the following:

1. The position of the perceiver – who is perceiving? Tourists, people with time on their hands? The public? From what location is one perceiving?
2. The experience of the five senses in the encounter of landscape – vision swells in importance as one perceives, but hearing, smell and touch can also influence evaluation of the landscape.
3. The discourse of landscape that "filters" one's sensory experience – "discourse materialized," as Schein describes, requires an attending to the socio-cultural ideas of landscape that influence our experience.[4]

Three philosophical approaches explain landscape perception as it relates to the landscape – perception as vision, perception as experience and perception as interpretation. Three different ideas of addressing a perceiver's position, sensory experience and discourse. I then add a fourth idea, perception as dialogue, and explicate each approach using the framework above.

PERCEPTION AS VISION

The central tension of the perception of landscape is between landscape as something we observe at a distance and landscape as something we

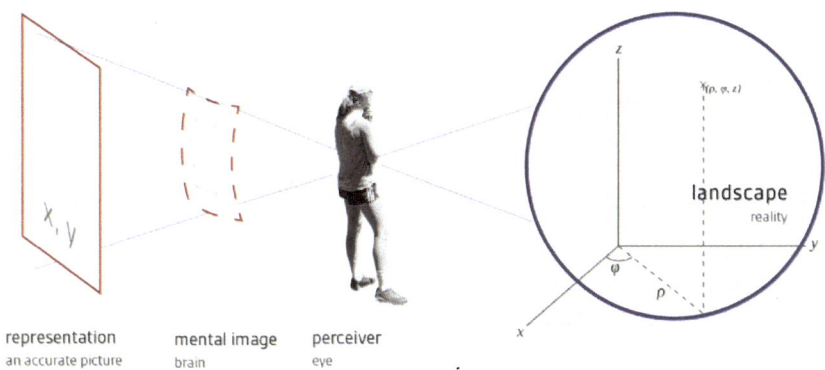

representation
an accurate picture
of reality

mental image
brain

perceiver
eye

3.3 Diagram of
perception as vision.

inhabit.[5] The first – landscape as a picture we observe – is the landscape of objectivity and modernism, of the ubiquitous industry of aerial photography, GIS and remote sensing. In visual preference surveys, the observer views the subject – the landscape – as a detached person who stands apart from the scene and thus can "truly" see it and assess its value. As the aristocratic, English estate owner walks onto the terrace of Stourhead, he surveys his property under his control as a "scene" – a pastoral tableau of grassy expanse, water and follies. As the GIS technician performs spatial analyses, she runs an algorithm based on land use zoning, vegetation extent, transportation corridors and assessor's parcels to generate a refined picture of the landscape to inform policy or development. Each observer looks from the outside or above, as if the landscape were separate from the person observing it. Perception as (solely) vision guides practices of the State and others in power who rely on what Michel de Certeau describes as the "God's eye view."[6]

Theory in support of landscape-as-picture suggests humans process visual images by receiving light stimulus in the rods and cones of our eyes, before using our brains to assemble a picture into something meaningful. According to the philosopher René Descartes, vision does not take place in the eyes, but in the operations of the mind.[7] The eye "sees" through the process of the brain, which builds an interior picture of an objective, material world. Meaning is not intrinsic to the landscape itself but is applied in the brain to pictorial elements. When someone sees a red barn, the brain first identifies the color and forms of light received, before linking the image with agricultural ideas of growth, food and the rhythms of the harvest.

3.4 Descartes' model of perception using the eyes and the mind.

The theory of landscape-as-picture has advantages. It is useful to take a photograph – from the earth or sky – and evaluate the image to determine the health of an ecosystem or its scenic quality. For instance, the percentage of pine trees weakened by the pine bark beetle can be measured through a spectral analysis of aerial photography. Vision is the most "scientific" of the senses, enhanced using cameras, sensors and different viewpoints. The camera frames what is seen, cuts out the

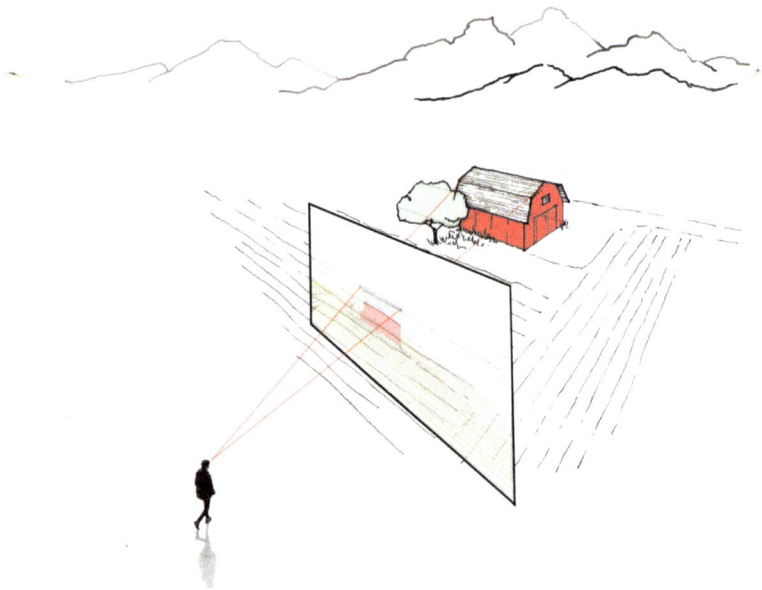

3.5 Diagram of an observation of a red barn, according to traditional perception where landscape is a visual image of reality.

messiness of reality and orders our attention. As Cosgrove eloquently states:

> Landscape is thus a way of seeing, a composition and structuring of the world, so that it may be appropriated by a detached, individual spectator to whom an illusion of order and control is offered through the composition of space according to the certainties of geometry.[8]

In this way, landscapes are measured, planned and produced through the assembly of information sent through the gauntlet of developers, designers, citizen feedback and city council meetings. Cartographies subsume place. The indignities of architectural business, the drumming up of projects and the limits of time, inform the limits of visual analyses.

A theory of landscape-as-picture also has disadvantages; it is too simplistic. The explanation of vision as "seeing" the landscape is limited due to the messiness of life as a multi-sensory, subjective and extensive milieu. The process of understanding the landscape (and ourselves) must include relations, networks and power imbalances, much of which is not visible. The review of landscape imagery is just one method in a suite of possible engagements.

PERCEPTION AS EXPERIENCE

Phenomenologists reject the idea of an objective view from outside of a place. People, including designers, inhabit a landscape from within, subject to environmental conditions, historic relations and daily rhythms. Phenomenology is the description of human experience from the first-person point of view. It is an acknowledgement that when the observer separates what is "out there" from what is "in here" one has already made a fatal distinction that obscures landscape and perception.[9] The singular position of the perceiver means the environment, the experience, cannot be generalized; it is not the experience of others. Landmark texts, such as Christopher Tilley's *The Materiality of Stone*[10] and Gaston Bachelard's *The Poetics of Space*,[11] do not attempt to explain the workings of the landscape, remaining open to any and other experiences.

Maurice Merleau-Ponty, the French philosopher, locates perception in the body and from within the body.[12] The body is a permanent part of one's perceptual field, as one touches things, sees objects, hears sound.

3.6 Diagram of perception as experience.

representation
writing, drawing

perceiver
within the landscape

Humans have a kinesthetic awareness of the world. We exhibit a spatial orientation towards actions or tasks, ways we might make use of objects or relate to others. Merleau-Ponty rejected the idea of vision arising from a detached, objective observer.[13] The landscape is not a neutral backdrop to the theater of human activity. Landscape is integral to humanity, not separated from the world.

The red barn in the agricultural fields is an encounter, not a picture. That is, the observer does not see a red barn and then think about how red barns are no longer utilitarian objects in the landscape. The observer's encounter with the red barn triggers immediate emotions and values as a part of the seeing, a preference or a meaningful engagement with the barn that may (or may not) be nostalgia. There is no distinction between seeing and thinking, body and mind, in perception as experience.

Often, phenomenology does not attempt to speculate about or explain experience, just describe.[14] It attempts neutrality of the eyewitness by shunning evaluation, preferring to point to the observer's own positionality in the landscape. In phenomenological writings, the use of first person over third person allows the reader to do the work of evaluation while acknowledging the limits of perception.[15] If there is a universal value or position behind perception, it comes from phenomenology's preference for local knowledge, for the perception of the farmer who maintains the red barn over the visitor who takes a picture of it. The positionality of a rooted individual brings lived experience to bear on the perception of a place, which gives them a deeper and more complex bodily knowledge (although see Tuan's essay on rootedness describing how locals' vision may be obscured).[16]

Immersion in the landscape is both the strength and weakness of the phenomenological approach. A phenomenologist experiences a landscape as a participant, more fully understanding a place and its constellation of relations. At the same time, the phenomenologist's understanding of a place has defined limits because of human limits and constraints. I may have a visceral reaction to the beauty of the windswept coasts, but I need the geologist to show me its dangers. I may celebrate the historic battlefield but need the Native American historian to see the battle as part of a larger system of oppression. Those writing on phenomenology at its height tended to be white and male, offering a view of the world from a distinct but narrow perspective in which many aspects of a place went unrecognized.[17] The phenomenologist worked to immerse himself in the landscape but it came at the expense of an honest reckoning with systems or structures of oppression.

PERCEPTION AS INTERPRETATION

If perception as experience relies only on the personal experience of those who write about it, the underlying structure of the landscape remains hidden. Scholars rooted in Marxism suggest experience is a fragile guide to the landscape, given its inability to interpret what cannot be seen.[18] They recognize that perception can be faulty, resting on a shaky foundation of assumptions, innuendo and the visual glance. Things we take as a natural organization of the world are not inherent. The landscape could be organized in many different alternative ways. Thus, perception may require a person from outside a landscape with an understanding of history, the structures of capitalism and political context to *interpret* what is seen. If the person perceives from within, they cannot sense in any perceptive way the surrounding structure and the distribution of power. Perception as interpretation moves the perceiver back outside of the landscape, not to look objectively at it, but to extend the interpretation of the space into other spaces, into the past and future.

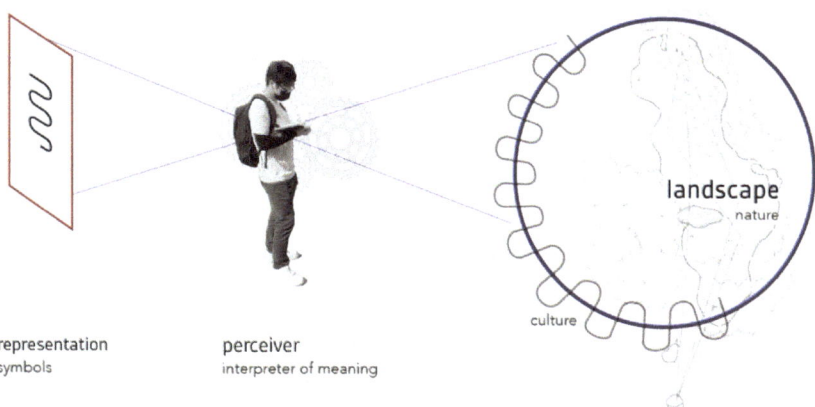

representation
symbols

perceiver
interpreter of meaning

landscape
nature

culture

3.7 Diagram of perception as interpretation.

Like perception as vision, perception as interpretation prioritizes the visual. Seeing becomes the predominate method of perceiving. However, this is not vision as the perception of light or an image, but vision to enable a reading of the landscape – an interpretation of symbolic elements that explain. The focus pivots to the landscape as a socially produced object. And if socially produced, then the perception of this landscape contains the potential for a better understanding of social relations ... relations materialized in the landscape.[19]

Metaphorically, landscape is like text. The eye sees black ink in contrast with the white page, the mind assembles these black shapes into letters and words and then interprets the meaning of these words based on past cultural conditioning. So then, the designer sees landscape elements and assembles them as sequences of meaning based on past design and ecological learning, as in Spirn's *Language of Landscape*.[20]

In this argument, landscape literacy is something that can be developed and practiced.[21] Designers develop an ecological mind or a cultural sensitivity with practice and training, but this specialized knowledge (of which this book takes part) does not mean other important knowledges should be ignored. Those who do not write about the landscape (or cannot write at all) may also have perceptive faculties attuned to different landscape elements – the threat of an approaching car, an unstable riverbank or a sidewalk where people are picked up for loitering. In research with unhoused people, their ability to move through the city or park, to access food and water and to keep themselves safe relies on a creative resourcefulness and flexibility different from that of domiciled residents.[22] Their interpretation of landscape is different as well.

Perception relies on an ability to interpret socio-cultural signs and symbols produced in the landscape according to one's ideology. Cosgrove describes the idea of landscape (which guides perception) as a "visual ideology."[23] To perceive the landscape is not limited to taking a photograph or sketching a space but incorporates an understanding of one's own ideas about space, as well as the ideas of others. This is landscape as signifier – landscape as "a linguistic expression of the complex cultural processes that mark the social evolution of the modern world."[24]

James and Nancy Duncan expounded on landscape as a text to be read.[25] Landscape-as-text requires a close reading to uncover ideological underpinnings. They argue for shared cultural interpretations of what they call the "naturalized" landscape; naturalized, here, meaning a condition of such prevalence that it is assumed to be "the way things are":

> If landscapes are texts which are read, interpreted according to an ingrained cultural framework of interpretation, if they are often read "inattentively" at a practical or nondiscursive level, then they may be inculcating their readers with a set of notions about how the society is organized: and their readers may be largely unaware of this. If, by being so tangible, so natural, so familiar, the landscape is unquestioned, then such concrete evidence about how society is organized can easily become seen as evidence of how it should or must be organized.[26]

It is this textual reading of the landscape that requires critical theory to uncover its meanings. It requires a person to "peel back the veil" of capitalist dominance. Traditional landscape perception – what we sense with our eyes and ears – becomes a response to a shallow experience of surfaces, underneath which lie the "true" relations or the way things are. This deeper cultural perception relies on education, a perception unavailable to those without a modicum of learning about the landscape. This is a common thread in *The Language of Landscape*, as Spirn laments the loss of the knowledge of landscape reading, of ecological clues, and thus, of the ability to construct more resilient places.[27] The signified requires one who can interpret accurately the nature of the sign.

Landscape symbols can both reveal and obscure. The creation of the gardens at Stourhead in England may be initially perceived as an assemblage of grass, pond, trees and that cozy white temple, but *interpreted* as Western symbols of power: sweeping lawn, Greek follies and pastoral scenery. Each landscape element has meaning. Each symbol indicates privilege to British citizens of the time (and arguably to present-day people in the West). Each symbol also obscures the process of landscape construction – the sources of wealth, the maintenance of the grounds, the pillaging of Greek statuary. The gardens of Stourhead were built by Henry Hoare II, the grandson of a goldsmith-banker who built his family's wealth from the extraction of gold by African laborers in places such as Guinea. Money in turn passed on to future generations to remodel and construct the grounds of Stourhead. Hoare designed the gardens himself, showing "a visual sensibility and imagination of a high order."[28] The wealth and prominence of the family relied on the transformation of (stolen) gold wealth into landscape symbol.

3.8 Stourhead estate and the perception of the English aristocrat.

The perceiver's position remains unresolved for perception-as-interpretation has shown that the perceiver must step outside the landscape to delve into the histories, policies and structures that shape the landscape. However, a perceiver is never "outside" the landscape. The designer remains immersed in the landscape which s/he evaluates. Furthermore, perception-as-interpretation postulates a hidden landscape that cannot be seen (or sensed) but it remains unclear how it can be seen/known. Duncan and others suggest critical theory can reveal but this is critical theory from a certain (academic) position of Marxism and now political ecology, a mostly inaccessible critical theory that posits a static landscape from a fixed viewpoint.

PERCEPTION AS DIALOGUE

Whether perception as an (outside) observer rooted in objectivity, as an insider experience rooted in subjectivity or as an interpreter of signs found in the landscape, our idea of how people perceive must explain our relationship with the landscape accurately or us to assess the landscape. I address unresolved problems with the prior modes of perception by proposing a model of perception based on dialogue, drawing from the work of Tim Ingold and Mikhail Bakhtin. The process of dialogic perception is the way people engage with and understand our relationship with the landscape.

Dialogic perception begins with an understanding of perception as something intertwined with the landscape (as in perception-as-experience), not as something separate. Here, I rely on the persuasive writings of Tim Ingold, an anthropologist from Scotland, who studied Laplanders in the arboreal regions of Norway and Sweden. As Ingold states: "We should cease thinking of perception as the computational activity of a mind within a body and regard it instead as the exploratory activity of the organism within its environment."[29] Rather than Descartes' model of the mind interpreting what the eye sees, Ingold's model begins with people moving around their environment and perceiving with their whole bodies, using all of their senses. As soon as we separate the senses into individual components, as soon as we distinguish between the person and the landscape one inhabits, we no longer accurately comprehend the process of perception.

The dialogic process includes both perceiving and *being perceived*. Ingold quotes the painter Marchand: "In a forest, I have felt many times that it was not I who was looking at the forest. On some days I have felt that it

3.9 Landscape dialogue uses all senses to experience textures, weather, water and place.

was the trees that were looking at me, that were speaking to me."[30] This is the reciprocity of perception, Berger's "reciprocal nature of vision."[31] Perception as dialogue requires an Other. This Other needs to be separate enough from the person to have different thoughts and ideas, but similar enough to be able to speak the same language of understanding, to be part of the same network. The metaphor of dialogue suggests an intertwined and poetic ephemeral alignment.

> The gaze is caught up in a dialogic, exploratory encounter between the perceiver and the world, in which every movement on the part of the perceiver is a questioning, and every reaction on the part of the perceived is a response.[32]

In landscape as dialogue, the subject perceives from within an ongoing relationship, not as an outside observer separate from the environment and not as an inside portion of a general milieu absorbed in the environment, but as a continuous companion, a fellow traveler in conversation with

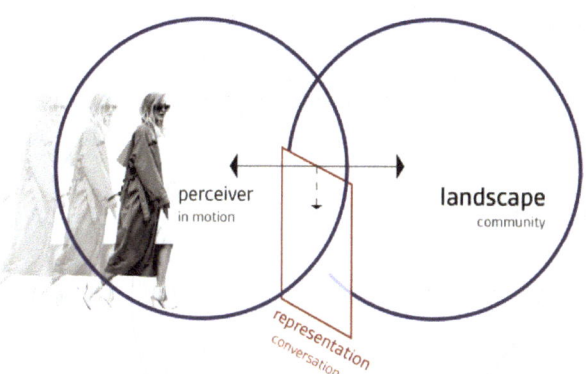

perceiver
in motion

landscape
community

representation
conversation

3.10 Diagram of perception as dialogue.

the landscape and other people. Designers cannot exit these intertwined relations to see things "objectively"; nor are designers unthinkingly reacting to environmental stimuli. It is the designer's position as both within and without the environment that makes it possible to dialogue.

As both within and without the landscape, the designer listens to the landscape as a voice to be heard, rather than reads the landscape as a text to be interpreted. The act of listening has unique qualities of embodied immersion and receptivity. A voice contains inflections and tones that express meanings beyond the literal words. It is difficult to sever the speaker from the speech, while it is easy to sever an author from the text. While speaking, we whisper to draw someone in or raise our voice to emphasize. We gesture with our hands. And the receiver of our wisdom, the hearer, participates in the act of dialogue through a receptivity to both verbal and non-verbal nuance. This can take the form of listening passively, as well as active forms of asking questions, raising the eyebrows, clarifying, arguing, lamenting, wondering why someone feels that way or responding with empathy.

In perception, the sense of hearing differs from seeing. The ear is an opening to the mind, while the eye filters immediate experience. It is the difference between text and the spoken word.[33] How people think of hearing corresponds more closely to perception as dialogue than how we think of vision:

When you see things that are far away, they are perceived to be at a distance, but when you hear far-off sounds they seem to be coming from a distance. The space of hearing, then, is not set over against you, the listener, but streams toward you and into you. It is a space not of place but of flows, where nothing can be divided and nothing measured. Your auditory experience is essentially participatory.[34]

The designer's position in relation to the landscape is one of participant, as an undivided, embodied experience. Through the movement of sound emission, we experience landscape's activity. We move through the world according to the rhythms of night and day, of tides and weather, of growth and decay. Perception uses the senses as part of a journey through space, an active engagement with a place.

This is not to say that vision takes a back seat in dialogue with the landscape. Vision can also be a reciprocal activity, as part of dialogue. In conversation, it is important to see a person's face, look into their eyes, monitor their expressions. Seeing works with and supports hearing as we perceive the world. Consider a red barn situated within a larger landscape with sounds of the highway permeating the background, the wind in a row of trees and a large combine in the distance plowing through a rectilinear field. Each landscape element contributes both aurally and visually to the scene or the experience.

Any model of perception must address difference – how do two different but reciprocal entities understand one another? This is the problem of *heteroglossia*, that a seemingly unified language of communication is really a multiplicity of languages with different meanings and ideologies. It is terribly easy to misunderstand another. Here, we turn from the anthropology of Ingold to the literary theory of Mikhail Bakhtin, the Russian philosopher of language.

Discourse requires a common language with at least some shared meaning to facilitate understanding. For Bakhtin, the problem of heteroglossia becomes reconciled, at least partially, through the activity of *dialogism* – different dialects within a single language communicating with one another (imperfectly). In Bakhtin's *The Dialogic Imagination*, he addresses heteroglossia using the lens of dialogue; people continuously respond to an ongoing conversation composed of traditions and prior works.[35] No conversation exists outside of the context of political, economic and spiritual life. Language speaks from and to previous works of language – previous perceptions and interpretations of a place. Dialogue is contextual. Difference then becomes part and parcel with an ongoing conversation to understand:

> Our utterance will in its very nature be dialogic: it is born as one voice in a dialogue that is already constituted; it cannot speak monologically, as the only voice, in some register isolated from all social, historical, and ideological contexts.[36]

Language contains within it the tensions of prior discourse. Utterances grope toward an answer or, at the very least, a temporary shared understanding.

Bakhtin would disagree with the unmediated nature of perception proposed by Ingold; each dialogic encounter is a mediation between word and object accounting for thousands of previous encounters in which that very word was said with a slightly different interpretation. Of course, Bakhtin is not addressing environmental experience but an experience of literature and conversation. Yet, like literary scholars, designers perceive the environment *in response to* an ongoing context of prior designs, ideas and cultural systems. An active understanding "indissolubly merged with the response, with a motivated agreement or disagreement ... Understanding comes to fruition only in the response. Understanding and response are dialectically merged and mutually condition each other; one is impossible without the other."[37] To understand the landscape, the claimed purpose of site analysis, is possible only through an active response, an engagement of practice of dialogue. Designers respond to the landscape through conversation; the "words" of place shift through times of perception, shifting from past meanings to present entanglements to future potential meanings.

Dialogue may be taken literally, not just metaphorically. Ojibwe peoples from North America dialogue with stones, trees and thunder. The animate speaking with the inanimate. Linguistically in Ojibwe, the word for stone gives the object a personhood, an animate character, or at least the potential for life. Portions of their traditional worldview were documented by the anthropologist Irving Hallowell in the middle of the twentieth century. Hallowell relates an anecdote about thunder:

An informant told me that many years before he was sitting in a tent one summer afternoon during a storm, together with an old man and his wife. There was one clap of thunder after another. Suddenly the old man turned to his wife and asked, "Did you hear what was said?" "No," she replied, "I didn't catch it." My informant, an acculturated Indian, told me he did not at first know what the old man and his wife referred to. It was, of course, the thunder. The old man thought that one of the Thunder Birds had said something to him. He was reacting to this sound in the same way as he would respond to a human being, whose words he did not understand.[38]

Hallowell attributes this dialogue to their metaphysical relations to the world, the way they see landscape populated with "persons" which colors their perception. Persons, integral to other-than-humans, are central to Ojibwe cause and effect, for how and why things happen. An Ojibwe is in constant dialogue.

All of us converse with and within the landscape. All of us are engaged in an ongoing dialogue with our environment, wherever we are at. It does not require a certain expertise or a certain type of place (i.e. wilderness) or a certain kind of relationship. However, we can practice becoming a better landscape conversationalist. The word "conversation" arises from the idea of familiarity with everyday conduct or behavior. This familiarity with a person or a place frees us to go deeper in dialogue to discuss meaning and detail. We can perceive the landscape with purpose.

DIALOGIC PERCEPTION IS MOVEMENT

Listening to another person moves beyond hearing a voice to an active attempt to understand what another person is saying. Listening requires an openness to others. It means we must be patient enough to attend. But that's not quite how the best listening works … for listening is a completely engaged and often displaced activity. We remove ourselves from our own position and *move* into a position of another. Psychologists promoting better communication discourage listening with the intent of immediately replying.[39] To listen, it behooves us to not generate a response until we first understand the other.

In our discussion of landscape dynamism (Chapter 1), we considered movement as something that happens within the landscape. Movement is also something that drives perception. A picture (as opposed to the experience) of a red barn is limited because it is static. Human perception of the landscape is dynamic; we perceive an unfolding scroll of reality. Alva Noë, the philosopher and cognitive scientist, describes perception as action; humans perceive by doing.[40] It is necessary to move around an object to determine what it is, its shape, its color, its consistency and wholeness. It is necessary to move within space to understand its qualities.

In particular, tacit understanding – the unspoken feeling of a place – depends on movement. Our understanding is based on past movements around an object, past explorations of other places. It is a sensorimotor perception of the world – an understanding of things and relations through movement through space.

We experience the world as unbounded and densely detailed because we do not inhabit a domain of visual snapshot-like fixations. When we hold our gaze fixed in that way, we do not look around, and insofar as we do not look around, we do not see. Vision is active; it is an active exploration of the world.[41]

Noë critiques the perception-as-vision approach by emphasizing the non-optical ways we experience embodiment in the world (reminiscent of Merleau-Ponty) with all of our senses.[42] He cites the example of taking off in a plane and how it appears that the nose of the plane rises up. In relation to one's body, the nose does not rise up, since the passenger is rising at the same time and in the same direction. The rising up then is not perceived visually; your balance, your inner ear (vestibular), detects it. Things that are "experienced visually depend on more than merely optical processes."[43] Moving forces us to deal with the body as separate from and yet related to other bodies, as the perceiver of the world and that which is perceived. There is no view of the world from "outside" or from the disembodied; we are living, breathing, moving participants in the landscape.

3.11 Moving while perceiving down Santa Cruz's Palisades Avenue.

Noë calls experience in motion "enactive perception."[44] And if movement, then presence. If we cannot be in place to listen, then we cannot perceive it (in a multi-dimensional way). Recall Descartes' vision of perception as occurring in the mind and contrast this with perception occurring in the body. An enactive perception suggests a closer connection between perception and the landscape. Ingold agrees, drawing from traditional knowledges of the landscape from various northern tribes of Scandinavia, Eskimos and the Ojibwe of the northern plains.

> For the Ojibwa ... knowledge does not lie in the accumulation of mental content. It is not by representing it in the mind that they get to know the world, but rather by moving around in their environment, whether in dreams or waking life, by watching, listening and feeling, actively seeking out the signs by which it is revealed. Experience here, amounts to a kind of sensory participation, a coupling of the movement of one's own awareness to the movement of aspects of the world. And the kind of knowledge it yields is not propositional, in the form of hypothetical statements or "beliefs" about the nature of reality, but personal – consisting of an intimate sensitivity to other ways of being, the particular movements, habits and temperaments that reveal each for what it is.[45]

Perception is not only passive receptivity to external stimuli. It is active. In the words of dialogue, the perceiver is not only listening but speaking. In the immediate sense, speaking can be in the form of drawing in the field, recalling a story about a place, constructing a path or discussing the landscape with another.

Speaking leads to a better understanding of place. We speak back to another: "Is this what you meant?" and "I hear you saying ..." A person wandering through a field of tulips does not just absorb the colored array but feels compelled to share their amazement with others. In the streets of the city, one speaks to others as one moves along the street, interrupting a train of thought to comment on the surroundings. One speaks to the place through movement, through choosing one path over another.

Each of these encounters and movements produces (in the language of social scientists) these spaces, gradually solidifying innumerable traces into the terrain of the city.[46] Like a desire line cutting a corner across the lawn, the daily rhythms of movement respond to environmental conditions but in turn merit a response from the landscape as it remakes itself to accommodate new paths and permutations. Even those without as much agency, like people experiencing homelessness, speak to the form of the city, inspiring policy and zoning measures that reconfigure sidewalks and streets, what is public and private and the opening and closing of businesses.

The designer, of course, speaks more specifically of the language of landscape and the design process – sketching observations, mapping places and making a spatial argument – a visual language of spatial measurements and material specifications. In the initial process of landscape dialogue, the designer speaks through sketching and trying out news ideas in response to the dynamic flows within and across the landscape. We now take a closer look at this important relationship between landscape, design and designer.

PERCEPTION AND DESIGN

As we dialogue with the landscape, we perceive the red barn, noting its color, texture and situation in the landscape; it is more a part of the landscape than part of our perception. Its nostalgic character, of a former way of life of the small, family farm, recalls the past, not as a distant memory, but as a present part of the landscape.

> To perceive the landscape is therefore to carry out an act of remembrance and remembering is not so much a matter of calling up an internal image, stored in the mind, as of engaging perpetually with an environment that is itself pregnant with the past.[47]

As we move closer to the barn, we listen with our senses to its presence, particularly if it is no longer functional. We assess its structure and condition, even if we are no architect. A cursory assessment of condition leads us back to nostalgia, to an understanding of the barn as perhaps more rooted in the landscape than if the barn was new or maintained. At this point, we might take a picture. We might categorize this barn as "picturesque." Whether consciously or subconsciously, the barn's situation speaks to us as part of a system of animal and food storage, of tourism and scenery and of color and light.

If, instead, we were to view a picture of a red barn, we might note its color and texture, we might perceive its condition, but fail to experience its situatedness in the landscape, the processes of its coming-into-being and the reason for its decay. Relying on pictures or remote sensing tells us something different. Distance can clarify, situate or obscure. Consider design perception as a nested set of scales. Early in my career, I worked for a wetland ecologist who at the outset of each new project would drive

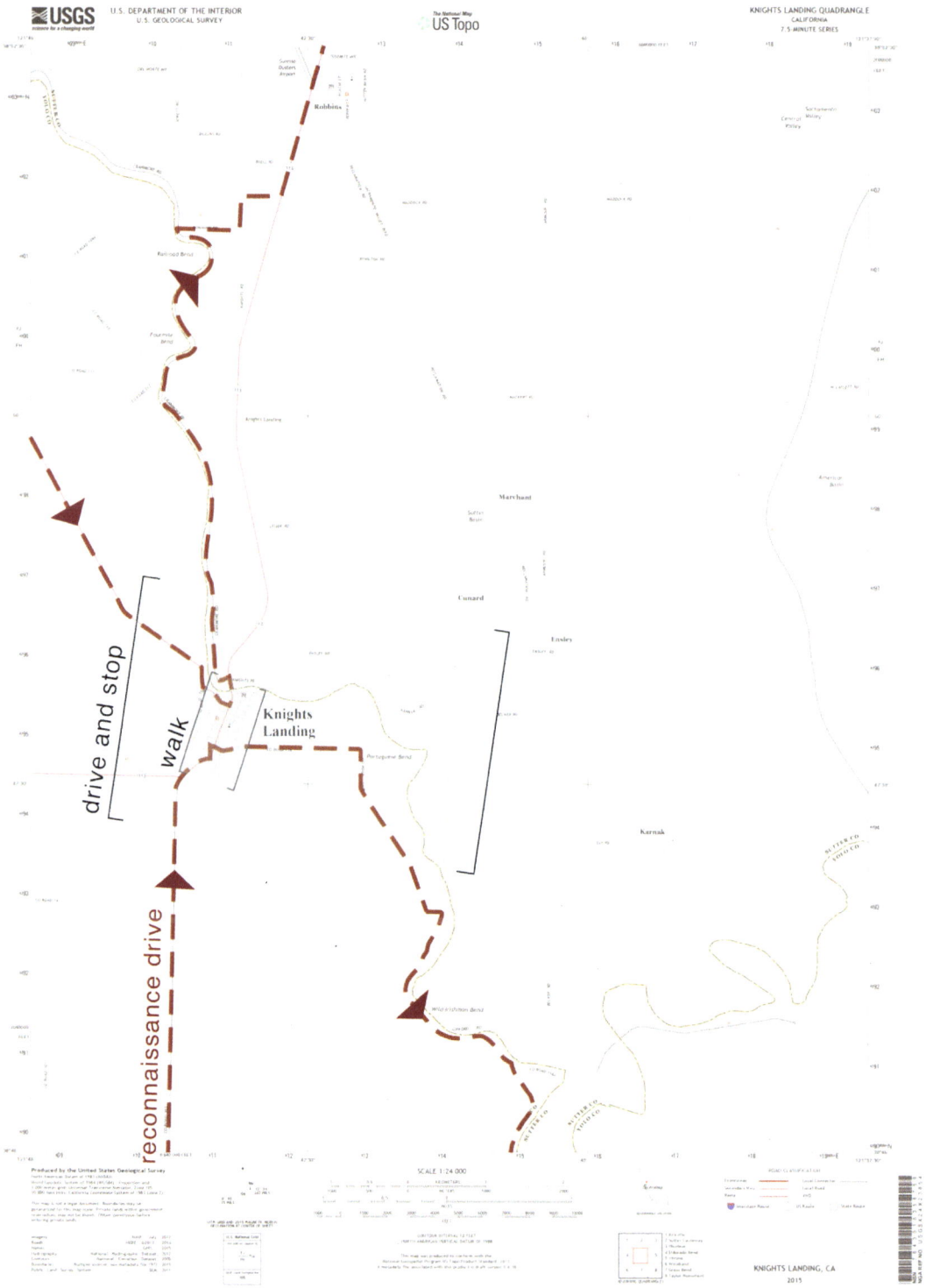

3.12 Nested movement/dialogue from the region to the site itself, in this case Knights Landing, California.

in a looping spiral that gradually got closer and closer to the site in what he called "landscape-level reconnaissance" (see Figure 3.12). For him, this reconnaissance meant getting to know a large area quickly through a windshield, a cursory process of exploration relying on moving through the landscape and connecting the dots. While we might spend most of our time on site evaluating a wetland, this division of landscape perception into scales or levels allowed us to move at different speeds depending on site proximity and detail of analysis.

Paynes Prairie case study

In the case of the Sweetwater Branch wetland design in Gainesville, Florida, we, as designers, shifted between different scales of landscape perception. Fifteen years ago, the United States Environmental Protection Agency notified Gainesville that their primary wastewater treatment plant was not treating wastewater sufficiently. In response, the city proposed a giant constructed wetland to "polish" the wastewater – tertiary treatment. The Sweetwater Branch of Paynes Prairie would be transformed from a constructed channel into three large wetlands that would treat (mostly clean) wastewater then disperse it onto the prairie. The proposed system of levees and wetland cells would have the added benefit of providing habitat for wildlife (alligators, birds, salamanders) proximate to Paynes Prairie that people could view from a network of nature centers, trails and interpretive exhibits we were to design at this urban–wild interface. So, even before visiting the landscape, we knew we needed to focus (our dialogic perception) on a few key things: the existing hydrology of Sweetwater Branch and the Alachua Sink, the historic character of the place, and the expansive nature of the landscape, in terms of views, feel and possibilities.

I was not familiar with the semi-tropical landscape of Florida. Arriving at Paynes Prairie, everything was new: the geology, hydrology, wildlife and the local community. This led to some fundamental questions: How should the designer approach a new project as an outsider? Is there an advantage to starting a new dialogue with a place, a "fresh set of eyes"? Or is the best designer the person who has a long-term dialogue with a place, as Wendell Berry would argue – a "care-giving" of landscape as a life work.[48]

The first field visit began with meetings, typical of design projects. Meeting the civil engineering group, the client team, the State Park staff of Paynes Prairie. Perception here requires keeping an open mind to people

3.13 Paynes Prairie State Park.

and their roles on the project. In a sense, avoid judging others too quickly (i.e. not all engineers think in black and white). We asked questions of the "native" hydrology – how did this system work 100 years ago? How does it work now? We asked questions of the people who manage the system and who might visit the system. Who were they and what did they do? What were they seeking?

Only then did we visit the site. Or at least the northern edge of the prairie near the site, the former site of Lake Alachua which appeared after heavy rains on the La Chua Ranch in the early 1870s, only to disappear 20 years later when a plug in the Alachua sink unexpectedly reopened, draining the lake. A park ranger led us on a tour of the existing paths and boardwalks, pointed out the wildlife – a snapping turtle, a roost of baby alligators – as we traced the water's flow to the sink. There was no red barn but there were a couple of extant farm buildings from the prairie's ranching days. As we stood by the buildings, the dialogue continued with the ranger, engineers and a local historian. We examined the Cracker-style architecture, particularly its weathering. Termites and tropical storms attack wood used in building in Florida. Weathering is an inescapable process, a process also in dialogue with a place and how it used to be and how it is becoming.

The next day, we had arranged to be by ourselves on the site of the future interpretive center, currently a swampy low area near the

3.14 Paynes Prairie ideas for viewing the landscape.

Sweetwater. The project manager had brought wading boots for us, so we put them on at the cars. The sounds of the highway receded as we walked into the swamp among the trees. She explained how to look for water moccasins as you wade through the ankle-deep water, using a healthy fear of poisonous snakes to heighten perception. Instead of the expansive place of the day before, we zeroed in on the specific topography and trees to locate a building and trailhead. We moved between the landscape experience and markings on the aerial photographs we brought with us, examining the micro-topography using knowledge of plant, soil and hydrology. Which trees, such as southern live oaks, should be preserved? Which trees, such as this clump of tallow trees, should be removed? We saw the play of sunlight on the ground and in the trees change during the day, incorporating shadow and light into potential ideas for interpretation. What was this place, with its long history and complex present of Seminoles, ranchers, university students and wildlife, trying to say?

3.15 Photo-rendition of interpretive center and garden.

Eventually, we designed a low-impact trail using crushed shell to reach a central interpretive center in the style of the wooden, Cracker ranch buildings. The open buildings will be cooled through large fans, instead of air-conditioning. Stormwater from the roofs and small parking lot/service entrance will charge an interpretive garden – a miniature model of the

3.16 Sweetwater Branch at Paynes Prairie boardwalk.

larger constructed wetland to be observed from the elevated boardwalk and interpretive center. The center will sit adjacent to the levee system where local visitors can already walk through a network of several miles of trail. Peeling off the levee, a wooden boardwalk transports the visitor into the wetland in a series of arcs that approach the wetland "islands," raised mounds of swamp tupelo and sweetgum. The sweeping boardwalks and trails immerse the visitor in the open and moving landscape inspired by the perception of openness from the site visits.

PRAXIS: THE SOUNDSCAPE

Much of the designer's task in dialogue with the landscape is visual – sketching, taking pictures, evaluating space. Here, I would like to suggest orienting landscape dialogue around sound. In this chapter, I argue that hearing is a metaphor for a holistic, sensory experience of the landscape. Hearing is also a literal activation of the senses. In the world of environmental psychology, the immersive experience of sound is called the "soundscape." Soundscapes are our acoustic environment., the sounds heard in place, experienced as a totality. The exploration of one's soundscape can lead to a more fully developed dialogue with the landscape.

Sound can contribute to well-being or detract from the experience of the landscape. A positive approach to sound includes an inventory and evaluation of pleasant sounds that enhance visitors' experience of the landscape. Design would then proceed to either protect existing sounds or even introduce new sounds into the environment. More commonly, a defensive approach is necessary to eliminate annoying or disturbing sounds.[49] For instance, at Paynes Prairie in Florida, city staff expressed concerns about noise from the highway interfering with the visitor experience of the reconstructed wetlands. We sited the interpretive center at an inconvenient distance from the arterial road, so visitors would have to walk into the space before experiencing it. The more distant location separated the visitor from automobile noise.

1. Sound quality

In measuring sound, use the ear and the decibel meter: the ear to assess the qualitative character of "noise" and the meter to assess the quantitative amount of sound. In landscape dialogue, the meter hears noise, but does not listen, while the ear listens but does not hear (at least in a way that can be directly measured).

For the ear ... visit a landscape. Take notes on the perceived sources of sound encountered on site.[50] Identify each sound and ask several questions of the sound, related to its source and duration:[51]

- Can the source of the sound be identified, i.e. a nearby highway? Alternatively, does the sound seem to be a general ambience, e.g. wind in the trees?
- How loud is the sound? This can be estimated by using a 5-point Likert scale ranging from 1 (very quiet/barely perceptible) to 5 (very loud).
- What is the duration of the sound? Is it continuous, rhythmic or an infrequent event? Continuous sounds include wind, traffic and utilities. Rhythmic sounds might include an overheard conversation or bird song. Episodic sounds include a train going by or construction equipment backing up. In addition, the sound may be temporary, unlikely to be repeated in the future, e.g. construction activity.
- Is it a positive or negative sound, in that does it contribute to or interfere with the intended purpose of landscape use?

If the landscape to be assessed is large, choose several key points, such as site entrance, interior or river's edge. At each point, assess the sounds by asking the qualitative questions of sound listed. Then, translate these notes to a map of the landscape. Locate specific sources of sound on the map and relate the key points to the sources.

2. Sound measurement

Using a meter ... while a decibel meter cannot distinguish sources of sound, it can give a general overview of sound levels across the landscape. Furthermore, since qualitative measurements of sound may not accurately reflect actual sound levels, a decibel meter can identify areas of the landscape that experience higher or lower sound levels than initially perceived. The process is straightforward. Establish a grid of points over a landscape using GIS or a surveyed map. Identify the coordinates in the field (using a smartphone, for instance). At each point, measure the sound level using a decibel meter. Link the sound level data to the coordinates by entering levels as point data into a GIS layer.

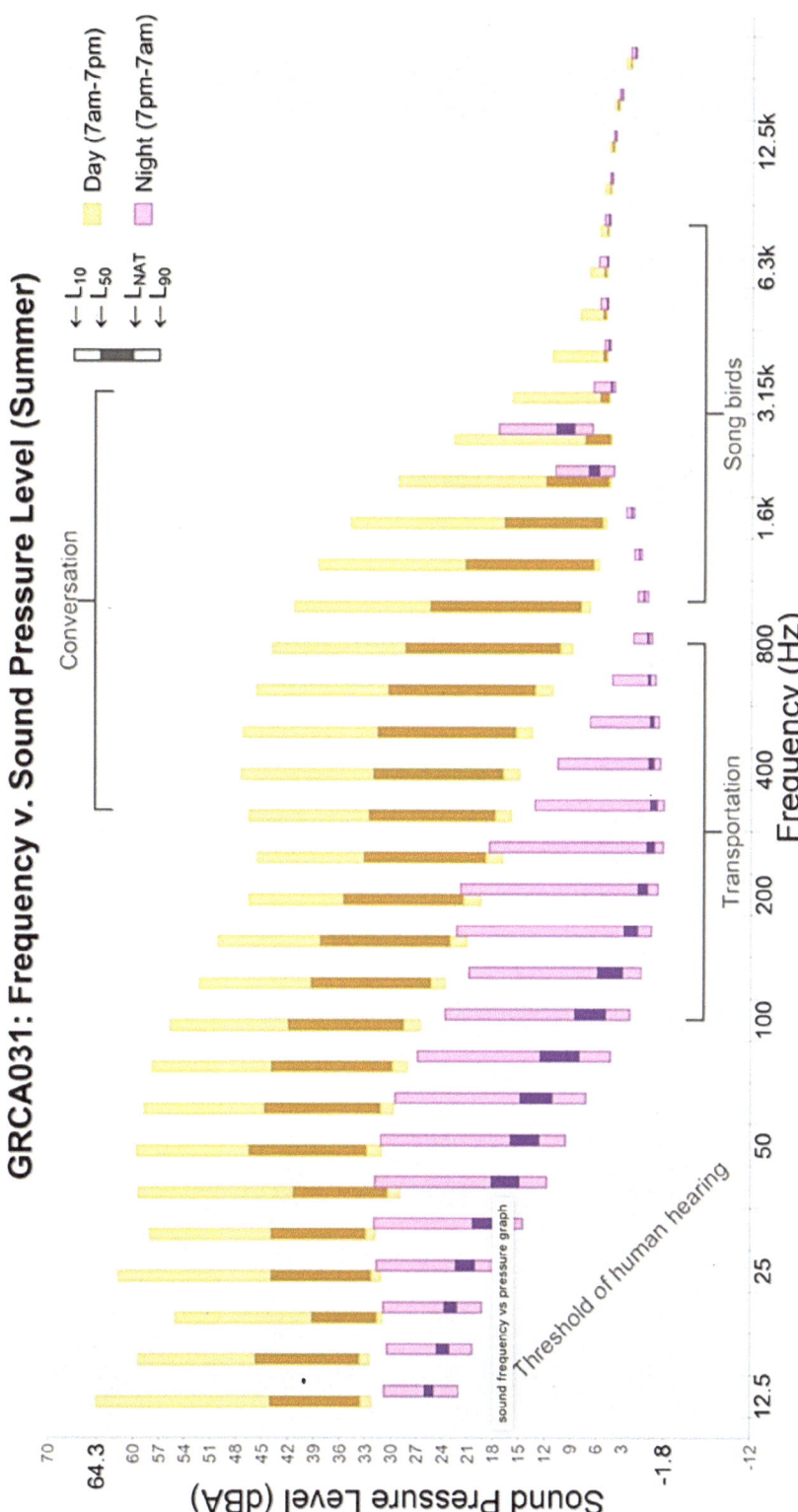

3.17 Frequency versus sound pressure level (summer) on the South Rim of the Grand Canyon. The graph shows that there are significantly higher sound levels during the day due to anthropogenic noise.

3.

Walking interviews ... as dialogue, the interview and its attendant structure, dynamics and anxieties occupy a central place in the social and design sciences. Inherently, though, an interview is dialogue, a series of language sounds exchanged to deepen meanings regarding a topic. If perception is primarily a motive activity, a movement through space, then interviews can be enhanced through movement. Thus, the walking interview – moving outside and into the landscape so that the environment informs the dialogue.

The walking interview has a niche history in the social sciences.[52] In assessing the landscape, the designer primarily engages in walking interviews with people who already know a place. For instance, in my own research on the homeless landscapes of Sacramento, I walked through important sites of informal encampments with a formerly homeless activist.[53] His rootedness in the social community and his inhabitance of many of the places under discussion provided spatial insight into the place.

In practice, the walking interview offers flexibility and rigor in spatial analysis, particularly in the relationship of pedestrian perceptions and movement through a space.[54] At the beginning of the interview, turn on GPS tracking on a smartphone to record the route and time during the interview, thus allowing for links to be made in the future between interviewee insights and urban spatial form. Following a "typical" route, as defined by the participant, walk through the landscape while discussing the places, experiences and stories, and the people they have encountered. Walking interviews, just like regular interviews, are enhanced by a careful preparation of questions beforehand related to the topics of most concern.[55] However, if the interviewee wants to talk about other aspects of the landscape, it is best to let them. After the interview, the designer can map the route and add important interview text in GIS as a series of points or hand-drawn on a map with links to the discussed landscape elements.

NOTES

1. Rachel Kaplan, Stephen Kaplan and Robert Ryan, *With People in Mind: Design and Management of Everyday Nature* (Washington, DC: Island Press, 1998).

2. Sally Schauman, "The Garden and the Red Barn: The Pervasive Pastoral and Its Environmental Consequences," *Journal of Aesthetics & Art Criticism* 56, no. 2 (Spring 1998): 181–190.

3. Schauman, "The Garden and the Red Barn," 189.

4. Richard H. Schein, "A Methodological Framework for Interpreting Ordinary Landscapes: Lexington, Kentucky's Courthouse Square," *Geographical Review* 99, no. 3 (July 1, 2009): 377–402, https://doi.org/10.1111/j.1931-0846.2009.tb00438.x.

5. John Wylie, *Landscape*, 1st edition (London: Routledge, 2007).

6. Michel de Certeau, *The Practice of Everyday Life*, trans. Steven Rendall, 2nd edition (1984; reprint, Berkeley, CA: University of California Press, 2002).

7. See René Descartes, *Descartes: Selected Philosophical Writings*, translated by John Cottingham, Robert Stoothoff and Dugald Murdoch (Cambridge: Cambridge University Press, 1988). For a discussion of Descartes' philosophy around vision, see Dalia Judovitz, "Vision, Representation, and Technology in Descartes," in *Modernity and the Hegemony of Vision*, ed. David Kleinberg-Levin (Berkeley, CA: The University of California Press, 1993), 63–86.

8. Denis Cosgrove, "Prospect, Perspective and the Evolution of the Landscape Idea," *Transactions of the Institute of British Geographers*, New Series, 10, no. 1 (January 1, 1985): 45–62, https://doi.org/10.2307/622249.

9. John Wylie, "Landscape and Phenomenology," in *The Routledge Companion to Landscape Studies*, ed. Peter Howard, Ian Thompson, Emma Waterton and Mick Atha (Oxford: Routledge, 2012), 72–83.

10. Christopher Tilley, *The Materiality of Stone: Explorations in Landscape Phenomenology* (Oxford: Routledge, 2020).

11. Gaston Bachelard, *The Poetics of Space* (Boston, MA: Beacon Press, 1994).

12. Maurice Merleau-Ponty, *Phenomenology of Perception* (Oxford: Routledge, 2013).

13. Maurice Merleau-Ponty, "Eye and Mind," in *The Primacy of Perception*, trans. C Dallery (Evanston, IL: Northwestern University Press, 1964), 159–182.

14. Wylie, "Landscape and Phenomenology."

15. John Wylie, "A Single Day's Walking: Narrating Self and Landscape on the Southwest Coast Path," *Transactions of the Institute of British Geographers* 30, no. 2 (June 1, 2005): 234–247, https://doi.org/10.1111/j.1475-5661.2005.00163.x.

16. Yi Fu Tuan, "Rootedness and Sense of Place," *Landscape* 24 (1980): 3–8.

17. Gillian Rose, *Feminism & Geography: The Limits of Geographical Knowledge* (Minneapolis, MN: University of Minnesota Press, 1993).

18. David Harvey, "The Sociological and Geographical Imaginations," *International Journal of Politics, Culture, and Society* 18, no. 3 (December 8, 2006): 211–255, https://doi.org/10.1007/s10767-006-9009-6.

19. J. Duncan and N. Duncan, "(Re)Reading the Landscape," *Environment and Planning D: Society and Space* 6, no. 2 (June 1, 1988): 117–126, https://doi.org/10.1068/d060117.

20. Anne Whiston Spirn, *Language of Landscape* (New Haven, CT: Yale University Press, 2000).

21. Peirce Lewis, "Learning from Looking: Geographic and Other Writing about the American Cultural Landscape," *American Quarterly* 35, no. 3 (1983): 242–261, https://doi.org/10.2307/2712650.

22. Cory Parker, "Bicycle Use and Accessibility among People Experiencing Homelessness in California Cities," *Journal of Transport Geography* 80 (October 1, 2019): 102542, https://doi.org/10.1016/j.jtrangeo.2019.102542.

23. Cosgrove, "Prospect, Perspective and the Evolution of the Landscape Idea."

24. Denis Cosgrove, "Modernity, Community and the Landscape Idea," *Journal of Material Culture* 11, no. 1–2 (July 1, 2006): 49–66, https://doi.org/10.1177/1359183506062992.

25. Duncan and Duncan, "(Re)Reading the Landscape."

26. Duncan and Duncan, "(Re)Reading the Landscape," 123.

27. Spirn, *Language of Landscape*.

28. Kenneth Woodbridge, "Henry Hoare's Paradise," *The Art Bulletin* 47, no. 1 (March 1, 1965): 83–116, https://doi.org/10.1080/00043079.1965.10788815.

29. Tim Ingold, *The Perception of the Environment: Essays on Livelihood, Dwelling and Skill* (London: Routledge, 2000), 260.

30. Charbonnier, 1959, p. 143, as quoted in Ingold, *The Perception of the Environment*, p. 276.

31. John Berger, *Ways of Seeing*, 1st edition (London: Penguin Books, 1990).

32. Ingold, *The Perception of the Environment*, 263.

33. Ingold, *The Perception of the Environment*, Chapter 25.

34. Ingold, *The Perception of the Environment*, 266.

35. M.M. Bakhtin, *The Dialogic Imagination: Four Essays* (Austin, TX: University of Texas Press, 2010).

36. Nasrullah Mambrol, "Key Theories of Mikhail Bakhtin," *Literary Theory and Criticism* (blog), January 24, 2018.

37. Bakhtin, *The Dialogic Imagination*, 282.

38. A. Irving Hallowell, "Ojibwa Ontology, Behavior and World View," in *Primitive Views of the World*, ed. Stanley Diamond (New York, NY: Columbia University Press, 1964), 49–82, at 34.

39. Marshall B. Rosenberg, *Nonviolent Communication: A Language of Life* (Encinitas, CA: Puddledancer Press, 2003).

40. Alva Noë, *Action in Perception, Representation and Mind* (Cambridge, MA: MIT Press, 2004).

41. Noë, *Action in Perception*, 72.

42. Noë, *Action in Perception*, 26.

43. Noë, *Action in Perception*, 26.

44. Noë, *Action in Perception*.

45. Ingold, "A Circumpolar Night's Dream," in *The Perception of the Environment*, 99.

46. Certeau, *The Practice of Everyday Life*.

47. Tim Ingold, "The Temporality of the Landscape," *World Archaeology* 25, no. 2 (October 1993): 152–174.

48. Wendell Berry, *The Unsettling of America: Culture & Agriculture*, 3rd edition (San Francisco, CA: Sierra Club Books, 1996).

49. J.C. Westman and J.R. Walters, "Noise and Stress: A Comprehensive Approach," *Environmental Health Perspectives* 41 (October 1981): 291–309, https://doi.org/10.1289/ehp.8141291.

50. For a comprehensive study of landscape and sound, see Jiang Liu, Jian Kang, Tao Luo, Holger Behm and Timothy Coppack, "Spatiotemporal Variability of Soundscapes in a Multiple Functional Urban Area," *Landscape and Urban Planning* 115 (July 1, 2013): 1–9, https://doi.org/10.1016/j.landurbplan.2013.03.008.

51. Note, some analyses of landscape sound divide sounds into three categories based on the source: (1) biophony – non-human sounds from organisms; (2) geophony – physical or non-biological sounds, usually from weather; and (3) anthrophony – sounds from human-made objects. See B.C. Pijanowski, L.J. Villanueva-Rivera, S.L. Dumyahn, A. Farina, B.L. Krause, B.M. Napoletano, N. Pieretti et al., "Soundscape Ecology: The Science of Sound in the Landscape," *BioScience* 61, no. 3 (2011): 203–216. I do not find these categories as useful, but they may help organize the assessment of sound.

52. James Evans and Phil Jones, "The Walking Interview: Methodology, Mobility and Place," *Applied Geography* 31, no. 2 (April 2011): 849–858, https://doi.org/10.1016/j.apgeog.2010.09.005; Phil Jones, G. Bunce, J. Evans, H. Gibbs and J.R. Hein, "Exploring Space and Place with Walking Interviews," *Journal of Research Practice* 4, no. 2 (November 14, 2008): Article D2.

53. Cory Parker, "Homeless Negotiations of Public Space in Two California Cities" (University of California, Davis, 2019), https://escholarship.org/uc/item/9x77627p.

54. Evans and Jones, "The Walking Interview."

55. Robert S. Weiss, *Learning from Strangers: The Art and Method of Qualitative Interview Studies*, 1st edition (New York, NY: Free Press, 1995).

CHAPTER 4

Landscape immersion: time in the landscape

An Inuit tribe in Alaska asked an architect to consider designing a cultural center that celebrated their northern culture. They arranged for him to visit the site they had in mind for the building. He arrived, met tribal leadership and they all went out to visit the place. As he talked with the group they moved to the center of the property and walked up a small rise. They stood together looking around. The architect talked of possibilities, about building locations, about his firm, until he noticed that a few of the others were walking away. Soon no one was left except a tribal elder, who also turned to go. The architect began to follow but the tribal elder gestured towards the landscape. Finally understanding, the architect sat down and stayed in that place by himself for an hour in silence … thinking, listening, considering.

The story is apocryphal. When I asked Johnpaul Jones, noted architect of the National Museum of the American Indian and the Southern Ute Cultural Center, if it really happened, he had no recollection of it. But when I first arrived at Jones & Jones as a young landscape architect, it was one of the first stories I heard. It was a compelling illustration of the firm's immersive approach.

Humans are active creatures … bustling, chatting, upturning stones. Movement enhances perception. But in movement, we can miss things. If landscape analysis is dialogue, most of us would rather be talking. We have big ideas and little time to communicate them. However, design is not just drawing, speaking and building. It is listening. This chapter argues that the quality of our design ideas rests on our ability to listen to a place, to immerse ourselves in the landscape. The practice of immersion is difficult to convey, like writing a book on meditation. Something is lost in the writing. Therefore, before I move to the more theoretical, I start with some concrete immersive learning techniques developed in, of all places, zoos.

DOI: 10.4324/9781003158943-5

It was 1976. Seattle's struggling Woodland Park Zoo had asked Grant Jones and a team of designers to offer a long-range plan for the zoo and a redesign of the gorilla exhibit, its signature attraction. Zoos in the first half of the twentieth century still clung to the Victorian idea of animal species to be catalogued, so wild animals were kept in cages with bars. Led by David Hancocks, the Woodland Park Zoo wanted to break the gorillas out of their concrete and metal cages.[1] The standard process of zoo exhibit design began with a visit to other zoos to see how they handled their gorillas. They skipped that step. Instead, Jones, Hancock and the design team set out to learn about gorillas in the wild – their needs, their resources, their habits. They called the process "landscape immersion." A team of zoo designers flew to the Democratic Republic of Congo and hiked through the jungle to experience the gorillas' home. Back at Woodland Park Zoo, their travels and landscape immersion inspired a gorilla exhibit based on the Central African bio-habitat, using plants native to the Congo. For the lowland gorillas, the benefits of an immersive habitat would be healthier, more sociable animals in a place closely approximating their habitat in the wild. For zoo visitors to understand the gorillas, they must immerse themselves in the gorillas' habitat as closely approximated on site.

4.1 Two caged lions at the Los Angeles Zoo, circa 1920.

This approach radically changed both how zoos cared for animals and how visitors learned about the animals and their habitat. Children began to make connections between the health of an animal and the health of an ecosystem. Parents began to question our conceptions of wildlife, zoos and the environment. More than that, visiting a zoo became an immersive experience. One of the designers, John Coe, put it best:

> To fully appreciate an animal (or plant) you must experience it while fully immersed in its natural habitat. The jungle-like plantings surrounding the approach pathways establish the exhibit context. Similar plantings within and surrounding the gorilla areas reinforce the conception that zoogoers are visitors in the gorilla's natural habitat.[2]

There is nothing native or "natural" about a zoo. Zoos have been slow to acknowledge that habitat for the largest animals who cover enormous territories in the wild cannot be reproduced in a confined area.[3] I begin a discussion of immersion with zoos because of their obvious synthetic nature as a human-produced landscape that attempts to incorporate natural processes into design. In a sense, all landscapes are synthetic. Immersion is not a spiritual experience of nature, but a process of engagement with the people/environment hybrid that is landscape.

Landscape immersion is a concentrated experience of a place. It organizes initial explorations of a "site" but also permeates subsequent design and construction. In each step of the design process, landscape immersion tells the designer about a place in a qualitatively different way than an aerial photo, map or photograph.

Although a step-by-step process is antithetical to subjective immersion, I offer the following preliminary suggestions:

1. Pay attention to how you arrive in a landscape – what are your own thoughts and feelings about the place? What are you bringing to this experience? What can be temporarily left behind?
2. Walk through the landscape, noting its spatial qualities. Where do you pass from one "room" to another? Sketch a rough map of these spaces.

Immersion gives a holistic, sensory experience of the landscape. It is an aesthetic experience open to all (although designers can develop immersive skills). In an effort to become more performance-based, the design disciplines have adopted some of the vocabulary and metrics of instrumental rationality (for instance, "performance metrics" at the Landscape Architecture Foundation). By using metrics, designers can assess the sustainability or

4.2 Woodland Park Zoo gorilla exhibit collage.

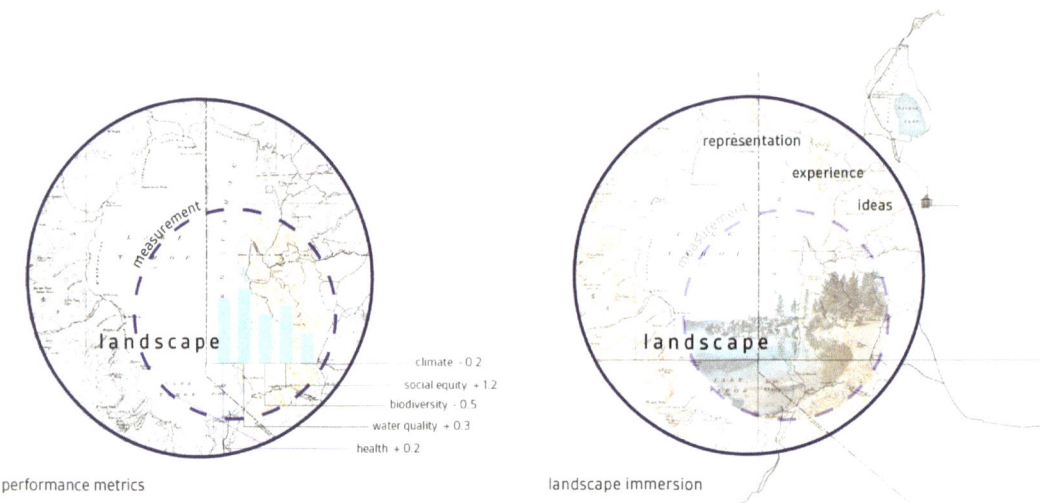

climate - 0.2
social equity + 1.2
biodiversity - 0.5
water quality + 0.3
health + 0.2

performance metrics

representation
experience
ideas

landscape immersion

4.3 Comparison of performance metrics to immersion.

popularity of finished projects, measuring success. However, metrics do not replace immersion. An assemblage of performance results does not account for the spatial, emotional and social qualities of being-in-the-landscape. Landscape immersion does. Its subjectivity is its strength for it paradoxically most closely mimics how other visitors experience a landscape.

Landscape immersion transcends disciplinary boundaries. By not dividing landscape elements into categories, an immersive approach retains an experiential view of the whole landscape. Specific, disciplinary knowledge may be used, as in the botany needed to understand which plants were to be used in the gorilla exhibit, but that knowledge works at first imperceptibly and then directly within a larger understanding of ecosystems and space.

AN IMMERSIVE APPROACH

Writing about immersion begins with a challenge. How do you describe a process that is so subjective? There might be as many ways of immersing in the landscape as there are designers? We have discussed the mechanics of information analysis. As we move through the landscape, we assess soil type, identify plant communities and record observations according to prescribed techniques specific to each discipline. We also get a feel for the place. This "feel for the place" cannot be quantified with consistency. Feelings are *verboten* in science. I knew a wetland ecologist who would not allow any of his employees to begin a sentence with "I feel that …" Ecology rests on rational, demonstrable measurements. However, the human species interacts with the environment in more complex and meaningful ways than a direct response to physical stimulus. What are these complex and meaningful ways and how might they inform landscape dialogue?

The work of Wilson Harris, in particular his novels, speaks to the difficult ambiguity of the landscape experience, the layers of meanings both imposed and inherent. Harris, hailing from British Guyana (now Guyana), surveyed the Cuyuni River basin of the country in his younger years and began writing poetry in an attempt to process the mystery of the landscape he was attempting to measure.[4] His commitments to the measurement of the landscape and the poetics of the landscape make him an ideal advocate for landscape dialogue, particularly as much of his writing laments colonialism's warping of Guyana's historic and ecological landscape. Harris interprets Guyana as the "land of waters":

> A great magical web born of the music of the elements is how one may respond perhaps to a detailed map of Guyana seen rotating in space with its numerous etched rivers, numerous lines and tributaries, interior rivers, coastal rivers, the arteries of God's spider … The spirit-bone of water that sings in the dense, interior rain forests is as invaluable a resource in the coastal savannahs which have long been subject to drought as to floodwaters that stretched like a sea from coastal river to coastal river yet remained unharnessed and wasted; subject also to the rapacity of moneylenders, miserable loans, inflated interest.[5]

For Harris that "music of the elements" becomes a "wordless music," not in the sense of without language but in the sense of an acknowledged gap between our understanding of a place and the mysterious place itself. The wordless music "brings an organ of animation" to a place, a "vulnerability" through which the strength of mutual dialogue arises – "omnipresence between characters—normally polarized and wholly separate—participating creatively and re-creatively in each other's fates and freedoms."[6]

Harris's writings can be difficult to understand. He used magical realism to interpret the landscape before which he found himself "mute" and "without expression." At times, a profound statement splinters into two or three meanings with alternative experiences now receding, now coming to the fore. But his commitment to engagement with the interior landscape of Guyana never changes. He could also be quite straightforward in his yearning to understand: "I and the leaves will always lie together. And know no parting." Harris' complex and rich experience of place inspired his attempts "to take the 'strangeness' of our shared world and build a language that could help us see and hear its multiplicity."[7]

4.4 Kaieteur Falls, Guyana.

We, as designers, evaluate the mysterious landscape as a hopeful and aspirational exercise. We acknowledge the uncategorizable nature of the landscape and the people within it through immersion. We then attempt a literacy with what we find, rather than what we would like to find. We notice things left behind, the meaning of presences and absences.

When one dreams one dreams alone. When one writes a book one is alone. The characters one re-creates may have died, or may have vanished into some other country, so one invokes them as "live absences," absences susceptible to being painted into life, sculpted into life, absences that may arise in carvings out of the ground, from dust, from the wood of a tree, the rain of a cloud: paintings and sculptures that are so mysteriously potent in one's book of dreams that they seem to paint one (as one paints them), to sculpt one (as one sculpts them), and in this mutual and phenomenal hollowness of self one and they become fossil stepping-stones into the mystery of inner space.[8]

Invoking the "live absences" of the landscape is not a static but a dynamic process that shapes the designer as it shapes place. It is more than time spent within. It is the cultivation and shifting of a landscape, the planting of a tree or the setting of pavers. We write notes, sketch images and speak through design. We cultivate ourselves.

4.5 Collage from Guyana, survey, water, land, ghosts.

Immersion from within and along

Others have struggled with the mystery of place but have focused on urban spaces. The French "little priest," Michel de Certeau, wrestled with humanity and city in a way that sparkles, digging into assumptions on what it is to see a place.[9] In the Western world, we assume knowledge of place comes from above, in the "God's eye view" of the world. Speaking as a visitor to the World Trade Center in New York City, Certeau saw "the gigantic mass … immobilized before [his] eyes."[10] There is pleasure in "seeing the 'whole' of looking down on, totalizing the most immoderate of human texts."[11] This totalizing vision attempts to escape the body and its limits. But the idea that the city is readable from the great height of a skyscraper is a fiction. The city is illegible, at least as a universal city, comprehensive and clear. Certeau defines the problem of illegibility as choosing between a false, totalizing view from above and the infinitely diverse views of the city as experienced on the street.[12] He deals with the problem by shifting from "what is seen" to "ways of seeing."

Landscape immersion means being on the ground, on the streets as opposed to in the air. It means wrestling with the lack of understanding of a city's processes, at least from a totalizing viewpoint. And it means a "way of seeing" as one moves through the landscape. The urban dweller of the streets sees the (illegible) landscape through movement:

> They walk – an elementary experience of this experience of the city; they are walkers, Wandersmänner, whose bodies follow the thicks and thins of the urban "text" they write without being able to read it. These practitioners make use of space that cannot be seen; their knowledge of them is as blind as that of lovers in each other's arms. The paths that correspond in this intertwining, unrecognized poem in which each body is an element signed by many others, elude legibility.[13]

In this urban sense, the landscape cannot be read. The intertwining paths disappear as soon as the walker passes. The spaces of the city remain obscure, disorienting the walker amidst the pluralities that defy observation. The urban landscape can only be lived, experienced and digested as a body. Certeau ultimately concludes neither the top-down or the street-level viewpoint provides a basis for interpreting the urban landscape by itself; rather interpretation requires a movement back and forth between a view from above and a view from below, the production of the powerful and the production of the powerless.

A woman stops here week-
ly and distributes food to
homeless people out of her car.
What does this mean to the
landscape, to the relationships
between landscape and people?

Urban traces of walking in informal spaces at
the fringes of automobility. Path is indicative
of need for pedestrian space leading over
bridge.

Street radius of 50 m allowing automobile
traffic to maintain speeds up to 35 mph,
despite the stop sign. There is no edge to the
asphalt, leading to gradual erosion of edge
condition as well as diffusion of automobile
space into the pedestrian space.

4.6 Intersection
of Highway 160
on-ramp and
Northgate Boulevard
in Sacramento,
California.

Immersion in the landscape does not mean reading a text, but engaging
in dialogue with the landscape, some of which we will understand and
some of which is illegible to us. We may recognize the spatial elements of
a landscape. We may not. The space we encounter remains the structure
upon which we hang the various landscape elements of inquiry, even if we
do not understand its influence and dynamism. Being in the landscape is
a subjective experience, from evaluating it as we move within and along
it, making comprehensive knowledge impossible to obtain.

Immersion takes time

Central to the experience of immersion is the idea of spending time. The
goal of immersive zoo exhibits, for instance, is to slow the visitor down
so learning can happen. While this is extremely difficult for a wiggly
toddler, it should be easier for adult designers. Pedestrian environments,

such as zoos, theme parks and small town business districts, are appealing because we can move around in a stimulating environment without fear of the danger of speed … a slow progression with stops. We can absorb the stimulation of rich environments. We need more time in the landscape, landscapes of specificity in which design occurs outside in the landscape. Landscape immersion blends the analytical with an openness to experience and the experienced.

We have discussed how perception of the landscape comes from movement through it. Moving within the landscape takes time in the landscape. It takes multiple visits, ideally to capture the changing seasons. As we inhabit a place, we learn of the place, an absorption of all that is there. But through dialogue. This is the central tension of immersion, the tension between the local person immersed in a home landscape and the expert visitor who visits a site. It is the tension between local folklore and generalized knowledge, a common debate in anthropological accounts.[14] Between the subjective and the objective. Immersion embraces both – time spent in the landscape, time living in place is invaluable, but also time learning how to see, hear and taste the landscape.

As time structures the linear nature of the written word, it also structures landscape dialogue. You talk, then I talk. Yet, the linear nature of time is not the only way of embracing the dialogic nature of the landscape. Wilson Harris, in his later novels, combined rich description of the landscape and the characters' interior lives, moving fluidly between the two. Time in flux. In a sequence, a character might die, often violently, only to re-emerge later in the story, as a living being (or maybe a ghost).[15] The past forces its way into the present, but displaced, always outside of one's grasp, so that observations are untrustworthy.

The sky was a dripping sponge over the river which had begun to swell. All the arid dusty watermarks of drought on the trees and bushes were disappearing, as if they, too, had been rubbed away leaving a clean but cracked slate, a web of broken lines nature had no desire to erase. These were salutary reminders of the displacement of the past, the basic untrustworthiness in every material image as well as in the conception of a supporting canvas. Were it otherwise the immaterial creation of freedom would have been banished forever, and no one would ever dream, in all perversity, of seeing it.[16]

Immersion as action

While everyone is immersed in environment all the time, we do not always attend to our surroundings. Immersion is both passive listening and active noticing. We position ourselves through slow movement to absorb the character of a place, to feel the wind on the skin, to note the sequence of landscape patterns as we walk through a space. We notice details. The kind of detail noticed (and sketched or noted) in the immersive process is not that a perennial is from the *Agave* genus, but that the vegetation is spiky. It is not that a tree is a *Quercus lobata*, but that its boughs create a canopy big enough to picnic under. Or even that the tree seems stressed. Immersion notes the spatial and functional details of landscape elements as a composite whole. Space and function unite in an aesthetic experience, so that the spiky plants, the oak canopy and the bare ground form a Mediterranean place in the shade. We do this as embodied people using the habits of design.

Immersion does not exclude measurement or rational thinking. Measurement yields insight into spatial form, relations between landscape processes (i.e. water) and landscape elements. Immersive measurement, measuring from within, is done with the body; the designer is primarily interested, in an immersive sense, in the human scale. As such, designers use their body as both a vessel of a certain dimensionality and as a reception of aesthetic experience. The designer walks over broken or smooth materials, stoops under a low threshold, paces a large open area, runs along a regional trail and gestures at the expansive view. Each of these is an embodied measure of the landscape experience, answering questions of comfort, size, movement and scenery. In a graduate studio in landscape architecture, students assessed a downtown from the perspective of a particular people group. One student chose the local homeless population. As part of a dialogue with the urban landscape she pulled a luggage cart and other belongings through a meandering transect, over curbs and out into the street, to experience the condition of having nowhere to store one's belongings. In addition to being informed of extra shelter beds downtown, she experienced a bewildering path up and over curbs, across driveways, around people and through parks. It was a bodily experience of urban movement.

Part of the immersive experience then is asking how this space is experienced by others, as we ourselves inhabit it for only a brief time. That may mean our immersion, our design meditation on the spatial qualities of a place, gets interrupted by other people. Talking to passers-by about

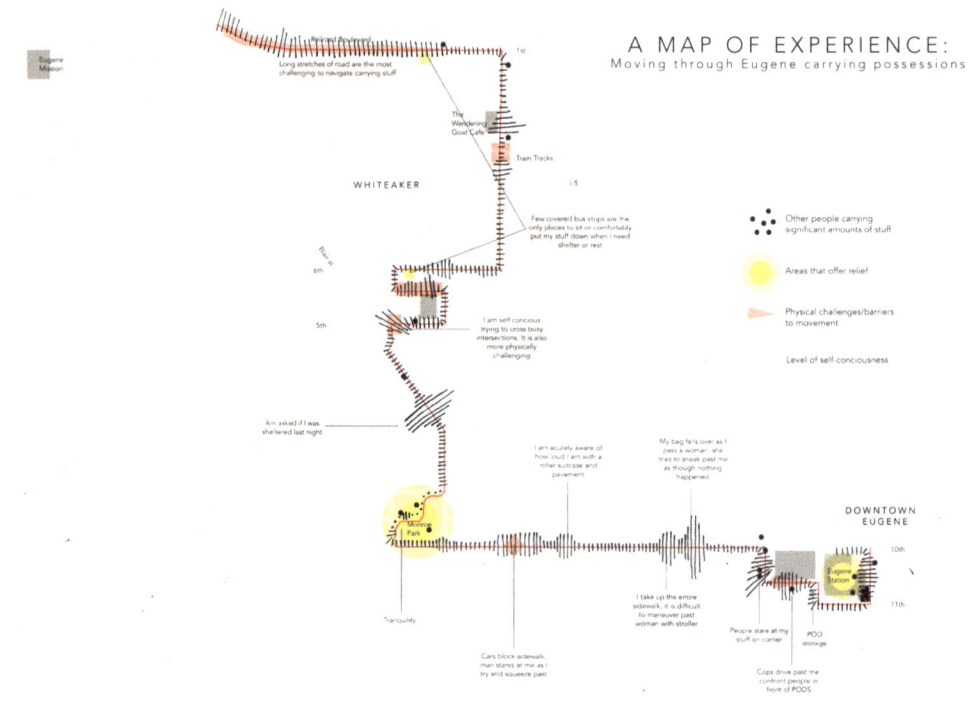

A MAP OF EXPERIENCE:
Moving through Eugene carrying possessions

4.7 A transect of a student pulling luggage through downtown Eugene, Oregon.

a place can yield information certainly, but more importantly, can get us out of our own heads, seeing a place from another perspective. We can miss details and their importance. Landscape dialogue assumes each individual has blind spots in the way they analyze a space and needs others to point these out. What is it we are taking for granted? How is our background and learning filtering our sensory experience of a place? As we visit multiple sites and multiple times, we build habits of attention to detail which can both make us better at evaluating a landscape or narrow our perspective of what is and what is not to be observed.

OFFICE AND FIELD

By prioritizing an experience of the outdoor landscape, I risk establishing another tenuous separation – that between the office and the field. So here, I discuss the similarities in the ways of thinking about the landscape and the differences in experiencing the landscape in both places.

The anthropologists, Gupta and Ferguson, describe the "field" as a tertiary mixture of "shifting entanglements of anthropological notions" of place and of an institutional study of a region.[17] The products of fieldwork in anthropology can relate more to the person doing the fieldwork, than the inhabitants of the field. Separation of the office (where one does analysis) from the field (where one collects data) leads to a problematic distinction between the worth and objectivity of the researcher and the perception of participants. In a similar way, the designer must acknowledge their own baggage they bring to the landscape (under analysis) as they immerse themselves in a place and a people (see Chapter 8). Our homes/backgrounds have a great deal of influence on the way we see a place.

If we perceive from within, relationally, it means the separation of "in the field" and "back in the office" may not be as clear as initially thought. Each place is a position, connected through the same relations central to understanding the landscape – client, resident, government, the flow of water, materials, etc. As we dialogue, we become aware of the effect of our position within the environment, within political systems and within spatial structures. My initial forays into the field during doctorate research took place in a neighboring town where I talked to people experiencing homelessness. I first drove to the area of interest, parked and completed my "field" work, before realizing I was creating an artificial separation between myself and those people and the place in which they lived. To combat this, I began riding the bus between towns, writing field notes as I got on the bus, observing and listening to the different people getting on and off as we moved through the landscape to the area of interest. Unhoused people frequented the bus as well, which led to a better understanding of the importance of transportation – its expense, its speed, its temporary bus communities formed for small durations, its rhythms and relation to stops on the way. I experienced some of the limits of homelessness and the trapped conditions of life in the midst of a car-centric culture. The field expanded from a single destination to a path of home to bus stop to walking route, until I reached the initial area of study.

Landscape architects often extoll the importance of contextual design,[18] of accounting for the broader processes in the landscape of hydrology, air, circulation and demographics. The collapse of the binaries home/ field and office/site suggest an embracing of context that, while not comprehensive, could better inform design flows. Designers do not turn off their perception, their senses, when moving between office and site. Perception is a constant and fluid experience.

That being said, the experience of a landscape is different when immersed within the landscape than when somewhere else. The immediacy of place cannot be replaced with Google Street View or mapping, no matter how

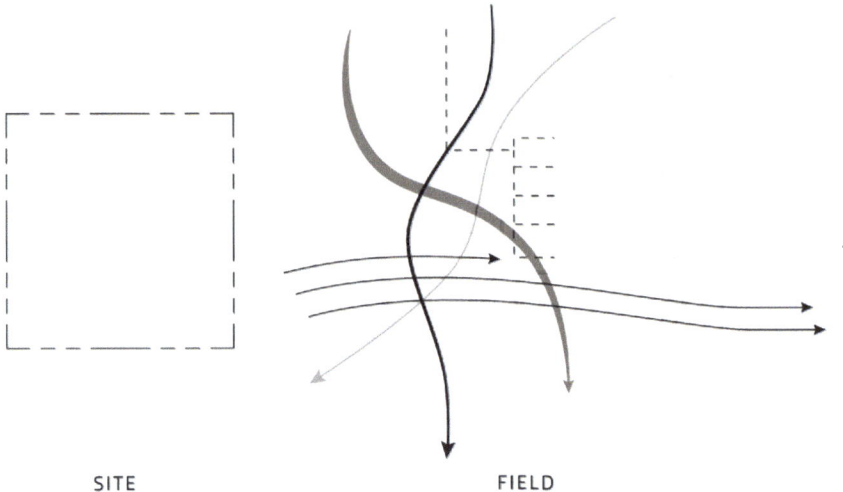

4.8 Diagram of field as a continuous space of flows.

SITE   FIELD

detailed. Design is both local and global; hence, requires attention to the micro-climate of place, while assessing the global life cycle of a material. To wrestle with the challenge of "site" and the interconnectedness of place, let us turn to some examples.

CASE STUDIES IN IMMERSIVE DIALOGUE

Lake Roosevelt National Recreation Area

Immersion may make sense for a small space, but it is less clear how it works in large-scale planning projects. To illustrate, we examine a national park in the United States. In a long-term effort to improve the recreational experience of Lake Roosevelt National Recreation Area, the National Park Service (NPS) hired our firm to complete a Shoreline Management Plan. As part of this effort, we spent three days on the lake and its shorelines. Lake Roosevelt is the 56-mile lake formed by the Grand Coulee Dam in Washington State, so the immersive experience/task was formidable due to the size of the park. In addition to circumnavigating the lake by vehicle, a park ranger motored us to remote campsites on an NPS patrol boat. At one point, we arrived at a campground, bounded out of the boat, began recording many of the challenges of the landscape, until I stopped the meeting to observe the boat had not been properly anchored and was floating away. The intrepid ranger shed much of his ranger uniform and swam out to the drifting boat to bring it back. True immersion.

We visited every campground and lake access point, took notes and photographs, met with operations staff and rangers. Gradually, data accumulated. Points of reference were discussed. And yet, we missed key challenges in the beginning. I had not "trained my imagination to go visiting."

There is a subtle beauty to the place, including historic buildings, grass-covered Palouse hillsides and pasture lands, but it is a National Recreation Area, not a National Park. It does not have that dynamic scenic viewpoint of a Mount Rainier National Park or unusual geologic features of a Yellowstone, and as such is used for active recreation. It is a water park. Boaters, fisher-people and jet skiers arrive every weekend from Spokane, Washington. People's second homes line the slopes above the lake. The "land"-scape component of the park sometimes extends only 50 meters from the shoreline on either side of the lake.

At first, as we interacted with park staff, the mission of the shoreline management plan was to add amenities to the campgrounds, a new boat launch, redesigning recreational spaces … physical changes. Soon, there was conflict between park staff who wanted to design new facilities and provide access and those who wanted to change policy, e.g. prohibiting campfires … changes in the ongoing monologue (for what is policy but prescribed words/regulations from those in power). Furthermore, private landowners adjacent to the park rose up to prevent physical changes,

4.9 Clockwise from top left: Grand Coulee Dam which impounds the lake; Lake Roosevelt; dock near Cedonia; second home above "private" dock; Hunter's Point recreation area; historic powder magazine at Fort Spokane.

new facilities requested by the broader public who lived farther way and who wanted to launch their boats from more convenient spaces. Landscape dialogue meant embracing tension at the points of conflict. The participants – planners, NPS staff and the public – were challenged to see the world as others saw it.

Our immersion in the landscape was less about sitting and absorbing the ethos of a place, it was about moving into and out of spaces, as recreationists do. The closest portion of the lake to Spokane is accessed via a dirt road. Driving the road, stopping at pull-outs to take photographs, passing locals in their trucks, and noting segments without gravel we realized the location would only accommodate a new boat launch if the NPS partnered with County government to improve the access road. And in public meetings, the local homeowners in the area of the proposed boat launch came out in opposition to public access (on public property). At this point, landscape dialogue suggests listening to angry voices; discussion/consensus is not possible until voices are heard (and questionable even then). But the NPS's informational approach of public meeting – transcribing meeting dialogue, proposal planning, publishing a report and public comment, and then another public meeting – does not lend itself to negotiation with locals. Information collection may be rigorous, but engagement is superficial. Many parks skirt this superficial process through long-term staff interactions with park neighbors, building a community of people and partnering with advocacy groups in a manner more consistent with an immersive approach. Immersive as relational.

At the second public meeting in Davenport, the closest town to the proposed boat launch, everyone knew what the issues were, what the proposal was and what people's positions were, so we set up the meeting differently. We rejected a one-microphone-model at a podium where individuals vented their frustrations to blank-faced park staff. We divided park staff and designers into pairs and stationed them by maps showing the park, zoomed in to specific controversial places where necessary. After a short presentation of plan and policy ideas, we spread out in the large conference room and talked to people individually, with one person taking notes and the other two discussing possibilities and pointing to the maps. This gave more people a chance to ask questions and propose new solutions, rather than just the angriest and most vocal of participants. Specific concerns arose out of the anti-boat-launch constituency that could be addressed, such as fires on the beach (in a very dry climate), litter and debris, the conditions of the road and the scenic quality of the park.

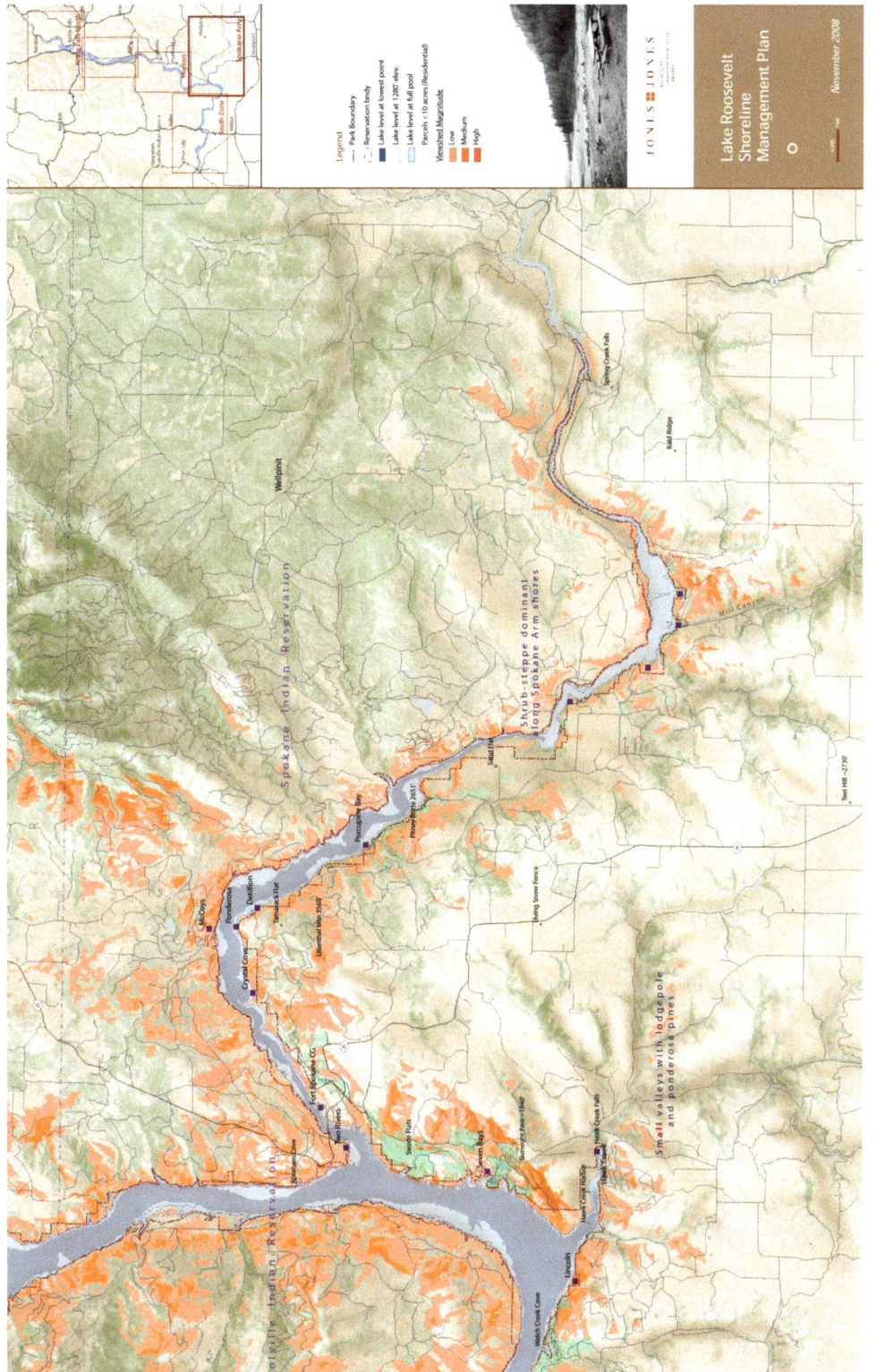

4.10 The Spokane Arm of the Lake Roosevelt National Recreation area showing viewshed. Orange indicates visibility from the water; darker orange means visible from more than one point. Lime green indicates areas of vacation homes.

Since the park consisted of the lake and the narrow shoreline, the tribes and other surrounding landowners controlled the scenic quality of the area. The park, like all parks, exists in a relational context with these communities, who were directly responsible for the aesthetic quality of the visitor experience. We set initial impressions and photographs of the hillsides within the context of a visual study in GIS. What could be seen from the lake and shoreline? And of those places, which were private, which were tribal reservation lands and which were subject to future development? When we brought representatives of the tribes, local government and the NPS together, we surrounded them with visual displays of our immersive experiences, maps and policy ideas. We suggested detailed ideas for improvements and detailed policy changes so park rangers had support for patrolling the park. It resulted in a Shoreline Management Plan that was both planning and policy, turning complications due to a lack of understanding of contextual issues to a focus on the immersive experience of the public, surrounding landowners and the working staff.

Seattle Chinese Garden

The landscapes of the NPS must rely on local or "native" inspiration for their materials and design. However, many design projects do not or cannot draw from the local landscape for inspiration. What does immersion look like for non-native landscapes? What does the designer immerse herself in? I turn to a long-term project that is a celebration of one culture amidst another. The intent behind the Seattle Chinese Garden is to strengthen social and economic relationships with Seattle's sister city of Chongqing, China. One city of one million in the United States and one city of 40 million in China (really a city-state). A visit of the Mayor of Seattle to Chongqing in the 1980s germinated the idea of a garden to bring the cities closer together.

Initially, the Seattle Chinese Garden Society identified over 20 sites in the city that could host the garden. The Society and city officials led a group of designers and politicians from the Chongqing Bureau of Municipal Greenery to visit the sites ... none of which showed promise. The site would need to host a traditional Sichuan Garden, a unique type of historic garden found on the Sichuan plateau of central China.[19] It would need to be hilly, more wild than other imported Chinese gardens modeled after the scholarly gardens of the Suzhou region. The South Seattle Community College asked the unwieldy team of designers to visit property adjacent to their campus. The designers from China immediately stated: "This is the place!" Landscape dialogue in 20 minutes. The site had a phalanx of trees to the south and west of the site, but more importantly, sweeping views of South Seattle and downtown. The scenic qualities of the site attracted the eye of the garden's major donor, Boeing Inc., who envisioned entertaining clients in one of the garden's pavilions as a client's new plane flew into the King County airport below.

4.11 Traditional Sichuan gardens; clockwise from top left: Shibaozhai along the Yangtze River, Garden in the Wenshu Monastery, Chengdu, Sichuan bamboo garden and pathway, Penjing Garden in the Wuhou Shrine, Chengdu.

The Society entered into an agreement with the Community College. And here, the real long-term dialogue with the landscape began – a soils report, a geotech analysis, wetland delineation and traffic study. Bad news came back from the wetland delineation report. The headwaters of a local stream began on the west side of the site. Covered by a reed canary grass monoculture, visitors to the site may have been excused for missing the wetland, but despite its degraded condition, the potential for future restoration cannot be paved over, even by a garden, so the size of the garden needed to be cut in half to protect the wetland with a buffer. Cutting the designed garden in half turned out to work well because the geotechnical engineers then discover over 10 meters of construction debris underneath the surface composing the "hill" the garden would descend. A digging into the history of the site uncovered an unscrupulous contractor who had built structures farther south on the campus and, wanting to avoid dump fees, decided to dispose of the debris on the site. Our joint Chongqing–Seattle design team was on the second iteration of the design of the buildings and squares and paths of the garden. Now, instead of a million dollars in funds going to the construction of the garden, it will have to go for earthwork – the scraping and stockpiling of topsoil, the removal of the unstable debris and the building back of a smaller hill upon which the Floating Clouds Pavilion would be placed.

Each step of the dialogue process, each setback, occurred in the midst of an immersive process of digging, the literal uncovering of problems and challenges. Immersion was not within a "native" landscape for the design to be inspired by local ecosystems, but an immersion in a non-native, synthetic landscape for the design to be informed by the structural and ecological qualities of the site. It began with an experience of views but then required digging before structures could be placed. At which point, the underground topography of the "good" or structural soil becomes the most important consideration in design. Immersion, for the designer, after the initial site visit, is a returning to the field with the geotech engineer, with the wetland ecologist to ask questions, probe unknowns and receive a clearer picture of the place. It is not the final report, but the probing process in the field, that transcends each discipline's basket of knowledge. Immersive dialogue is conversing on site about the place and our roles within it.

4.12 Initial Seattle Chinese Garden drawing from the Chongqing Municipal Bureau of Parks and Greenery.

Immersion for the Seattle Chinese Garden did not just take place on site. A traditional Sichuan garden requires four elements: rock, plants, water and buildings. Centuries of plant-collecting in Sichuan province have distributed native Sichuan plants throughout the world's nurseries: the peonies and the tree ferns of China, readily available in the United States. It is the rock that is different. The Pacific Northwest of the United States is composed of granite and volcanic rock from the Cascade Mountains. Sichuan gardens use limestone. So, we teamed with the lead landscape architect of Chongqing to tour rock yards, informal quarries and stone outcrops to come up with a local stone with the same characteristics as that of Chongqing area.

Despite the setbacks, the Society eventually constructed an entry promenade and the first formal space, the Knowing the Spring Court. We, as designers, switched from the immersive point of view trying to understand the space to the immersive point of view of the visitor trying to experience another culture. Design calls for an imaginative immersion into a future space – how will someone enter and move through this space? How will they understand the symbolism of plants and architecture? In this case, gardens in China would never be considered a "site"; they make use of existing stone, plants, water and borrowed views. We designed a garden that would seem like it had always been there … using natural features of the place, from the "discovered" creek to the rocky outcrop overlooking Seattle.

4.13 Seattle Chinese Garden's Knowing the Spring Court before opening.

PRAXIS

Immersion is the least linear practice. However, it is helpful to think of landscape immersion as a sequence of arrival, meditation and departure, primarily because time within a landscape will have a beginning, middle and end. The designer understands that important processes and elements do not inhabit the same timeframe as a brief site visit; however, traces of those processes can be seen in the landscape. Ideally, there can be a series of immersive experiences at different times of the day, which recognizes that the immersive experience itself is not sequential, in the sense of having to do something first before completing the next thing. Embrace the ambiguity and detail of immersion.

1. Movement and arrival

The first immersive "technique" is movement to and into a space. Circle a site in a gradually diminishing spiral. The movement of the car, bus,

4.14 Photo-sequence of arrival to Cesar Chavez Park, Sacramento, California.

bicycle through the surrounding spaces speaks of differences in access, power and relationships. What changes as one gets closer to the landscape? How do differences in the direction of approach affect the experience of arrival? The arrival is critical from the perspective of the designer as conversationalist and as a person. Arrive at a site with openness and humility. Shed client expectations (temporarily). Photograph the arrival sequence(s). Get out of the car and walk into the space from the existing "entrance" and any arrival spaces that may be designed in the future. Take pictures every 20 meters and label them on a map. If time allows, sketch this sequence instead of using the camera.

2. Landscape meditation

After arriving within the landscape, meditate for several minutes with eyes closed and ears open. Find a space away from others where you will not be bothered and just sit. There is no practical purpose here, other than letting go of purpose, however temporarily. This is not always possible. In one initial site visit for a large wildlife rescue center, I arrived late due to a delayed plane flight. I spent the now shortened site visit in dialogue with the client. The literal dialogue was important, but I was frustrated at not getting to know the landscape. So, after returning to his offices for meetings the rest of the day, I made an unscheduled visit back to the landscape at dusk to wander through the spaces and spend time quietly listening to the wind in the oaks. (Unfortunately, I also brushed against poison oak in the dark – using my body as a measurement of human–plant relationships.)

In urban areas, a very simple experiment uses the body and social norms to better understand a place. Stand in place for five minutes. Do not move around, lean against a wall or take notes. Just stand and observe. If someone engages in conversation, feel free to speak but return to your standing in silence when done. In the presence of other people, this will be uncomfortable. People are used to being in space with a purpose, and subtly evaluating others based on their behavior (whether striding, drinking coffee, or asking for a handout). After five minutes, find a new spot and stand in place for ten minutes. Then find a spot to take notes about your experience afterwards. How did you feel? Was this a welcoming space?

In contrast to the solitary architect standing on a hill described at the beginning of the chapter, most immersion in the landscape will be more walking and exploring than meditation. Movement should be slow with frequent pauses. This is particularly true for large landscapes. Our analysis of Lake Roosevelt National Recreation Area was challenging due to many informal camping areas only accessible by water. We relied on a boat to reach the shoreline, arriving at the beach and pushing up on foot into the riparian habitat. We marked "camps," as well as garbage and debris. Traces of camping told us how boat campers used the landscape. We climbed the side slopes adjacent to the beach, until we could see each cove's relationship with the beach, lake views and the dry watersheds. An engagement with access points and spaces beyond informs future design possibilities. Apparently, 98 percent of the visitors to Yellowstone never venture more than 100 meters from a road or parking lot. Designers need to be the 2 percent of the population who move into the interior. Engage with a place by clambering over and through it. Grab the sagebrush and pull yourself up. Immersion is active.

Sketch from different viewpoints. Perhaps the most immersive of the activities, sketching forces the designer to dialogue with a place. Use a book on sketching the landscape, such as Janet Swailes' studies of the archaeology of Scotland and Wales in *Field Sketching and the Experience of Landscape*. Sketching as an immersive activity is not necessarily an artistic activity (although it can be). The point of this drawing is not to get published in a book showing your sketching brilliance, nor is it to post to Instagram. Be as messy as possible if that makes the pencil flow across the page.

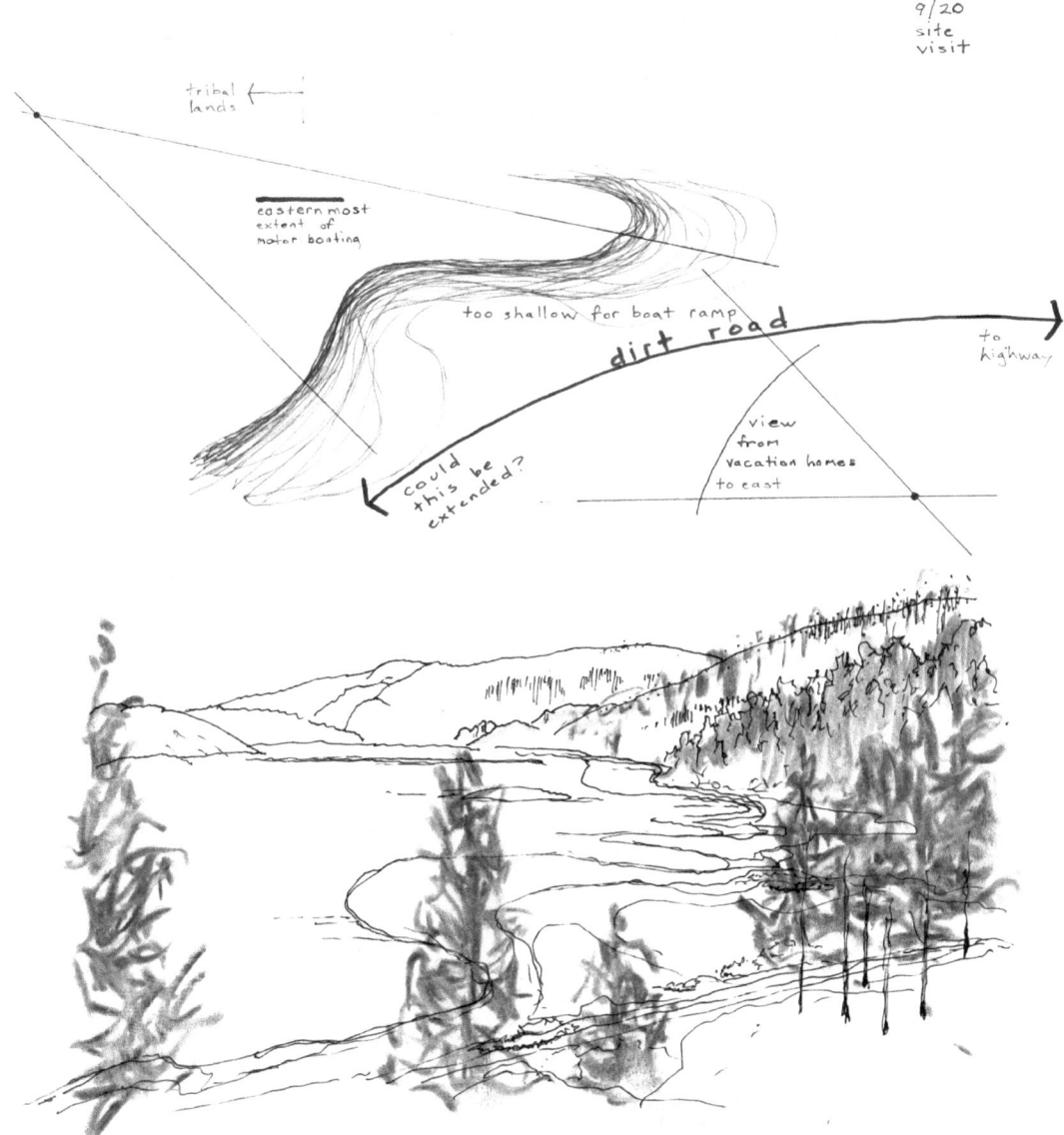

4.15 Sketching of a Lake Roosevelt cove showing key issues here: access and boat draft.

3. Departure

At the end of the visit, practice a leaving ritual before moving on to the next place. Field ecologists spend time cleaning their tools, surveyors inventory their equipment and save their data. The transition from one place to another should embrace movement and acknowledge landscape difference. Examples of leaving rituals include:

- Sketching a key map linking field notes, sketches and diagrams to an overall plan.
- Gathering and organizing found materials and sketches.
- Talking with a colleague about overall impressions.
- Taking a moment to yourself to re-experience the path you took through the landscape.

Develop your own transition ritual. Then write about the landscape experience immediately. After field research on homelessness in the urban landscape, I would immediately find a nearby coffee shop to compile notes and write down immediate impressions. There, I would get everything in my head into a notebook or laptop. Waiting too long leaves space for other things to intrude into your short-term memory and crowd out your experience.

4. In the office – what to do when immersion is not an option

While there are certainly many things a designer can do after working in the field when one gets back to the office, I want to address those times when an immersive experience is not possible. What should the designer do if a site visit is too brief or not practical? The short and best answer is to hire a local consultant from the area of interest who can visit multiple times and dialogue with the designer about the place. Absent this opportunity, here are some recommendations:

1. Every designer should be practicing immersion in whatever area they reside. Stand, notice, record and sketch in areas similar to the distant site. Become familiar with the process of immersion until it is a habit. Then, when one delves into the remotely available data on a place, it will be its own experience with more than one dimension.
2. Talk to people who live there. A friend or colleague can do a video walk-through of the landscape, narrating their own experience.

3. View photography and aerial photos. Open up Google Street View to "travel" through the space. The designer can then assimilate the habits of immersion into the information gathering process.

If no one can go to the site and practice immersion, larger issues of who should be designing, the energy needed for distant travel and the voice of local people complicate the design process. Find local partners.

NOTES

1. David Hancocks, *A Different Nature: The Paradoxical World of Zoos and Their Uncertain Future*, 1st edition (Berkeley, CA: University of California Press, 2001).
2. John Coe, "The Genesis of Habitat Immersion in Gorilla Exhibits Woodland Park Zoological Garden and Zoo Atlanta – 1978–1988," *(Unpublished)*, 2006, 2. See www.johncoe.com.
3. Hancocks, *A Different Nature*.
4. Fred D'Aguiar, interview with Wilson Harris, *Bomb Magazine*, January 1, 2003, bombmagazine.org/articles/wilson-harris/.
5. From "A Note on the Genesis of the Guyana Quartet," Wilson Harris. As quoted in D'Aguiar, interview with Wilson Harris.
6. D'Aguiar, interview with Wilson Harris.
7. Gemma Robinson, "Wilson Harris, New World Writer," *Stabroeck News*, In the Diaspora, March 19, 2018.
8. From *The Four Banks of the River of Space*. As quoted in D'Aguiar, interview with Wilson Harris, 2003.
9. Michel de Certeau, *The Practice of Everyday Life*, translated by Steven Rendall, 2nd edition (Berkeley, CA: 1984; repr., University of California Press, 2002).
10. Certeau, *The Practice of Everyday Life*, 91.
11. Certeau, *The Practice of Everyday Life*, 92.
12. Ian Buchanan, *Michel de Certeau: Cultural Theorist*, 1st edition (London and Thousand Oaks, CA: SAGE, 2001).
13. Certeau, *The Practice of Everyday Life*, 158.
14. Kent C. Ryden, *Mapping the Invisible Landscape: Folklore, Writing, and the Sense of Place* (Iowa City, IA: University of Iowa Press, 1993).
15. Wilson Harris, *The Guyana Quartet* (London and Boston, MA: Wilson Harris, 1985).
16. Harris, *The Guyana Quartet*, 446.
17. Akhil Gupta and James Ferguson (eds.), *Anthropological Locations: Boundaries and Grounds of a Field Science* (Berkeley, CA: University of California Press, 1997).

18. For examples see Krista L. Schneider, *The Paris-Lexington Road: Community-Based Planning and Context Sensitive Highway Design*, 2003, https://trid.trb.org/view/756506; James A. LaGro, *Site Analysis: Informing Context-Sensitive and Sustainable Site Planning and Design*, 3rd edition (Hoboken, NJ: Wiley, 2013), http://site.ebrary.com/lib/ucdavis/Doc?id=10653568.

19. Jerome Silbergeld, "Beyond Suzhou: Region and Memory in the Gardens of Sichuan," *The Art Bulletin* 86, no. 2 (June 1, 2004): 207–227, https://doi.org/10.2307/3177415.

CHAPTER 5

Landscape relations: how connections structure dialogue

U.S. Highway 93 running north–south through the Flathead Indian Reservation had become one of the most dangerous roads in the nation. Fast Montana cars sitting behind a slow-moving RV heading north on U.S. 93 would become impatient, swerve out into the oncoming lanes to pass and would be hit by a local resident turning right on to the highway. Looking left they would never see the speeding vehicles coming from the right. Horrific head-on collisions.

In the 1980s and 1990s, deaths from car accidents increased. Drivers were also hitting large animals: deer, elk, black bear, grizzly bear. A car accident with an elk or a bear can kill the animal, but the driver and car do not come out too good either. The Montana Department of Transportation decided something must be done. They proposed a new design for U.S. 93. The new highway would get people traveling from Missoula to Flathead Lake and Glacier National Park to the north through the reservation as quickly as possible. It would be changed from two to four lanes the entire way, allowing people to pass slow motor homes. They would reduce the number of small roads and driveways accessing the highway that made it so dangerous. They considered vehicular movement in terms of speed and safety. They neglected to consider the Salish and Kootenai people, their history and their antagonism towards people driving through their landscape as fast as possible.

If we think of landscape as a static entity *unrelated to ourselves*, we will treat the landscape, its people and its animals as an obstacle to be traversed. If landscape is an object to be observed from a distance, we will not consider the impacts of our interventions (despite environmental regulations requiring such consideration). But landscape is inherently

DOI: 10.4324/9781003158943-6

relational. Nothing exists, especially a ribbon of pavement, in isolation. This is a truism, backed by religions, academic theory and quantum mechanics (or how the world works) – that the world is better understood as a network of relations, rather than a collection of objects.[1] Yet we treat the people and the landscape of each place as if they exist independently, as if their relations to the landscape and others has been severed. One method of severing some relational connections and processes is to construct a road through it. I describe a highway design project in Montana to explicate the relational aspects of design – relationships between people and people, people and landscape – that must be acknowledged and embraced in landscape dialogue.

Ninety-five percent of U.S. Highway 93 between Evaro and Polson sits within the Flathead Indian Reservation. The Salish and Kootenai tribal members who occupy the reservation arise from three distinct areas – the Salish from the fertile Bitterroot Valley to the south, the Pend d'Oreille from the west (present-day Idaho) and the Kootenai

5.1 The Flathead Indian Reservation with the Mission Mountains in the background.

from the valleys between the Rockies and the Pacific Coastal range in the north. Sham government treaties, the 1864 Gold Rush and white settlers/squatters forced the Salish within the Bitterroot Valley north to the newly established Flathead Indian Reservation in the late 1800s.[2] Soon, even that reservation was not "reserved" as the United States in the Allotment Act of 1887 opened reservation lands not designated as fee simple ownership to white settlers. Following a long legal battle with the tribes, white settlers flooded into the lowlands of the reservation, choosing the prime farmland along the valley bottom. The settlement was concomitant with the building of another railroad running from Missoula to Flathead Lake, guaranteeing white farmers access to Chicago and large food markets. Access inspires development. The railroad and its history became a symbol for the tribes of the oppressive occupation of their land by people who thought differently, farmed differently and lived within the valley differently.

Given the role of transportation in the opening up of the reservation to colonizers, solving the dangers of U.S. 93 by proposing *more* transportation infrastructure – the privileging of faster cars – did not seem viable. Yet, that is what the Department of Transportation proposed. In the design process, transportation engineers rely on "instrumental rationality" – a philosophy of maximizing efficiency. It is the optimization of a logical end, in this case the speed of automobile travel, whatever the means (disregarding moral and social considerations). In Western cultures, the application of instrumental rationality is technical, a reliance on quantifications to assess and determine economies and thus efficiencies. Landscape architecture and architecture use instrumental rationality as well, particularly in a world looking for numeric assurances. The disciplines choose an end – sustainability, public health or social equity – and take measurements to assess whether that end has been achieved. It informs performance metrics such as used by the Landscape Architecture Foundation.[3] Yet the approach has limits. It works to solve technical problems once identified but struggles to identify the problems in the first place. Problems are assumed as shared and inviolate (e.g. people not able to drive as fast as they would like). The Montana Department of Transportation (MDT) diagnosed the initial problem – people were dying in car accidents – but could not see larger problems related to the history of the people and landscape and how they relate to each other. For MDT engineers, the singular problem of car accidents could only be solved by the singular solution of a wider road. Other solutions, such as reducing speeds, reducing traffic, creating a different kind of road or travel by a different transportation mode, did not come to mind. By pulling things apart (road pavement, access, towns, forest) and dealing with landscape elements separately, engineering solutions missed the relational qualities of the landscape.

The Confederation of Salish and Kootenai Tribes sued MDT. As a sovereign nation, the tribes appealed to the Federal Highway Administration (FHWA) and generated an alternative solution of a two-lane road to improve the highway without making it wider and faster, and thus, less sensitive to the place itself.[4] The litigation dragged on. No improvements were made to the road. In the 2000s, the Confederated Salish and Kootenai Tribes, recognizing that eventually they would lose their legal battle, sought compromise. They negotiated a Memorandum of Understanding (MoU) between themselves, MDT and FHWA, that would allow highway improvements to move forward with specific cultural and physical conditions negotiated by the tribes.

The MoU brought together two distinctly different philosophies of transportation: one in which cars (and thus people) move through a space in isolation from the surroundings and one in which cars move in relation to their surroundings and others. Movement composes a huge percentage of public space in the United States. By ceding this function and space to engineers, the design/architecture community has neglected the daily rhythms of residents. The transportation engineers pursued a

5.2 White cross signifying fatal car accident along Montana highway.

wide, efficient highway centered on driving fast, as if these drivers and their speeding vehicles had no relation to the places they moved through. In later design discussions, I brought up the request of the residents of the small town of St. Ignatius to reduce the speed limit of the highway as it went through their town. The engineers in the conference room became angry. "Speed limits can only be changed by an act of the state legislature!" The state legislature represents Montana drivers for whom driving at high speeds was seen to be a human right; the town's request was ignored.[5] Calling attention to an assumed norm of speed upset a speed sensitivity. If highway speeds degraded social, cultural and public health and if small towns had a say on the highways that ran by them, then instrumental rationality as a foundation for decision-making could not be rationalized.

In contrast, the Confederated Salish and Kootenai Tribes were under no illusions about the importance of relationships. The Salish, Kootenai and Pend d'Oreille Indians have distinct tribal origins, histories and values but a shared perception of the landscape as a web of interconnected relations. If one relationship suffers, others suffer as well. This applies to present relations, as well as past: "Our history is written within our unique and specific cultural landscapes. These places hold the memories of our ancestors, speak to us in the present, and are crucial to our survival, as Indian people, into the future."[6]

Major infrastructure projects come into being and are maintained by and through a web of relations. These relations are uneven. They are framed by those in power to maintain dominance over those without. The railroad of the late 1800s brought white settlers into the valley, counter to the treaty, disrupting reservation dynamics – political dynamics that carry over to today, in the form of suspicious Native Americans and insecure white settlers. "Neutral," "rational" or "objective" highway improvements do not exist. Each additional lane, each added road cut and retaining wall changes the relationship of infrastructure–driver–resident. As well as wildlife.

The relationship between the Salish and Kootenai and animals is crucial to their life world. These relations define traditions, daily life, seasonal movement and tribal identity.[7] Animal "geniuses" who can speak and call on the power of song dramatically shape the landscape, providing the tribes a place to call home.[8] Encounters with wildlife in the landscape remind tribal members of ancestors and origin stories. Wildlife and humans are not two separate entities; they share kinship with the Spirit of Place. Animals tell people of seasonal cycles, warn of environmental damage and portend future events. Debra Magpie Earling, an author and

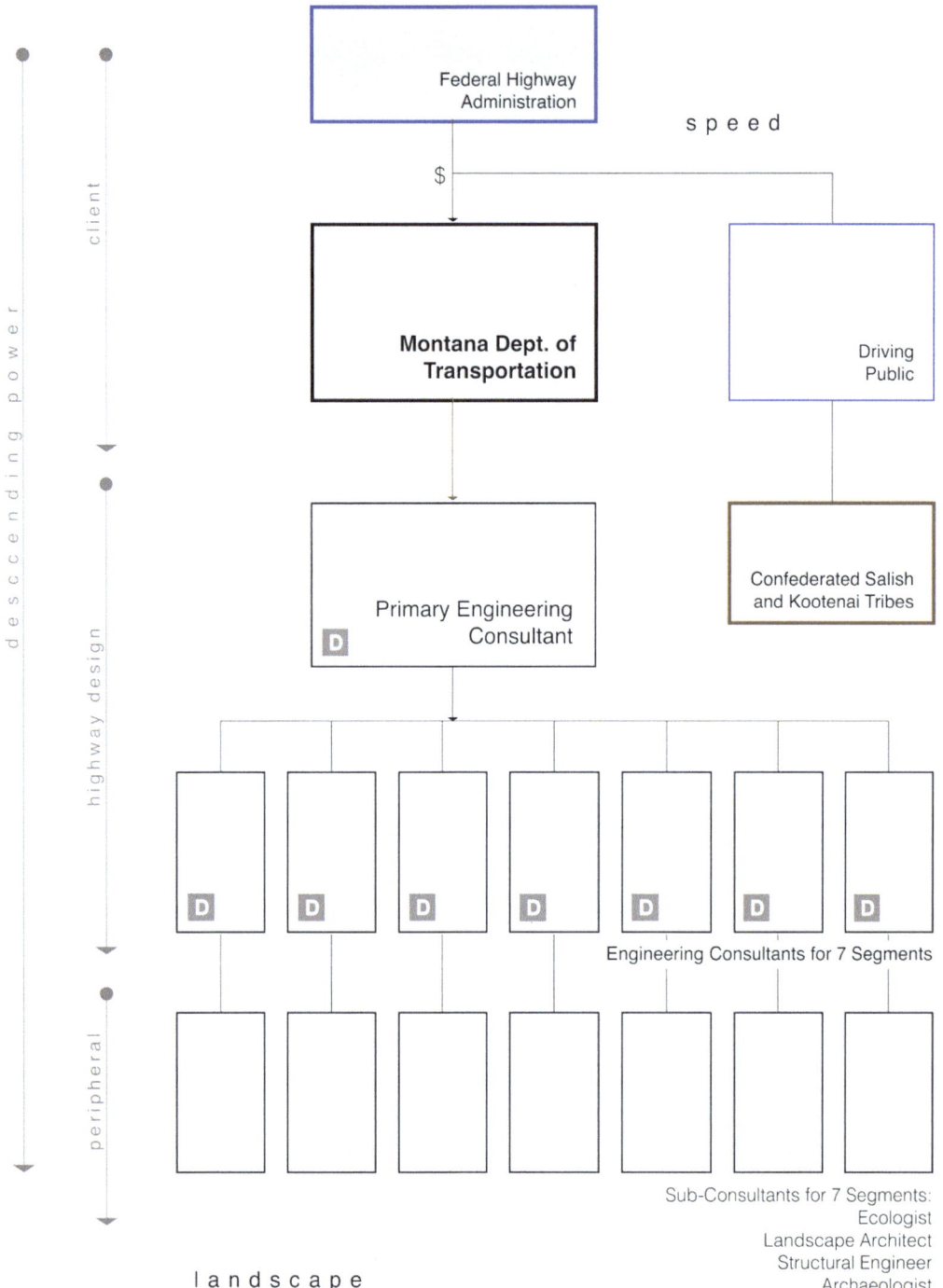

5.3 Diagram of initial power relations amongst U.S. 93 stakeholders – traditional engineering approach to highway design.

Confederated Salish and Kootenai tribal member, illustrates an example of this in a story told by an elder:

> If Rabbit appears suddenly and twitches his ears, we know to drop to the ground before lightning sizzles past us. Hummingbird turns red before he enters our lodges to signal a young person will sicken. Rocks soften beneath our feet when we pray.[9]

In addition to, and as part of, this close relationship, wildlife provide sustenance to the tribes through hunting, fishing and just being present. The tribes have dedicated wildlife and ethnobotanical staff to manage tribal relations with movement and habitat of animals. How then could the tribes' ongoing dialogue with their landscape be applied to a highway, that iconic symbol of twentieth-century rural America?

At the beginning of the MoU process, the tribes asked MDT to hire our firm to be an advocate for the tribes and for the landscape, based on our work on Paris Pike, a highway through the culturally significant blue-grass landscape of Kentucky. We spoke the language of landscape, engineering and design. The first process in the office was to visit the landscape, talk to the tribes and identify aspects of the landscape that needed advocacy. A highway was going to be "improved," whether the tribes wanted improvements or not, so could the highway be used, as Grant Jones put it, to "knit the landscape back together again."

The tribes initially proposed two wildlife crossings at Evaro and Ninepipes in the early 1990s as part of a U.S. 93 scoping session.[10] It was not until Julie Neff, a landscape designer at Jones & Jones who had been to Banff National Park where several major wildlife crossings allowed elk and moose to cross the TransCanada Highway in safety, proposed a network of crossings that a larger idea took hold. She dug into the design and location of those crossings, contacted Anthony Clevenger from Banff who was monitoring those crossings to determine their efficacy, and then generated presentation boards that showed how they might be implemented on U.S. 93. During the next meeting on the project, designers put the wildlife crossing presentation boards up on the wall, to generate discussions around the idea. As the boards were placed in the meeting hall, an engineer from the lead design firm came along behind and began tearing them down. A physical battle of two philosophies. Eventually, someone from FHWA said "No, leave them up. Let's hear what they have to say." And dialogue began.

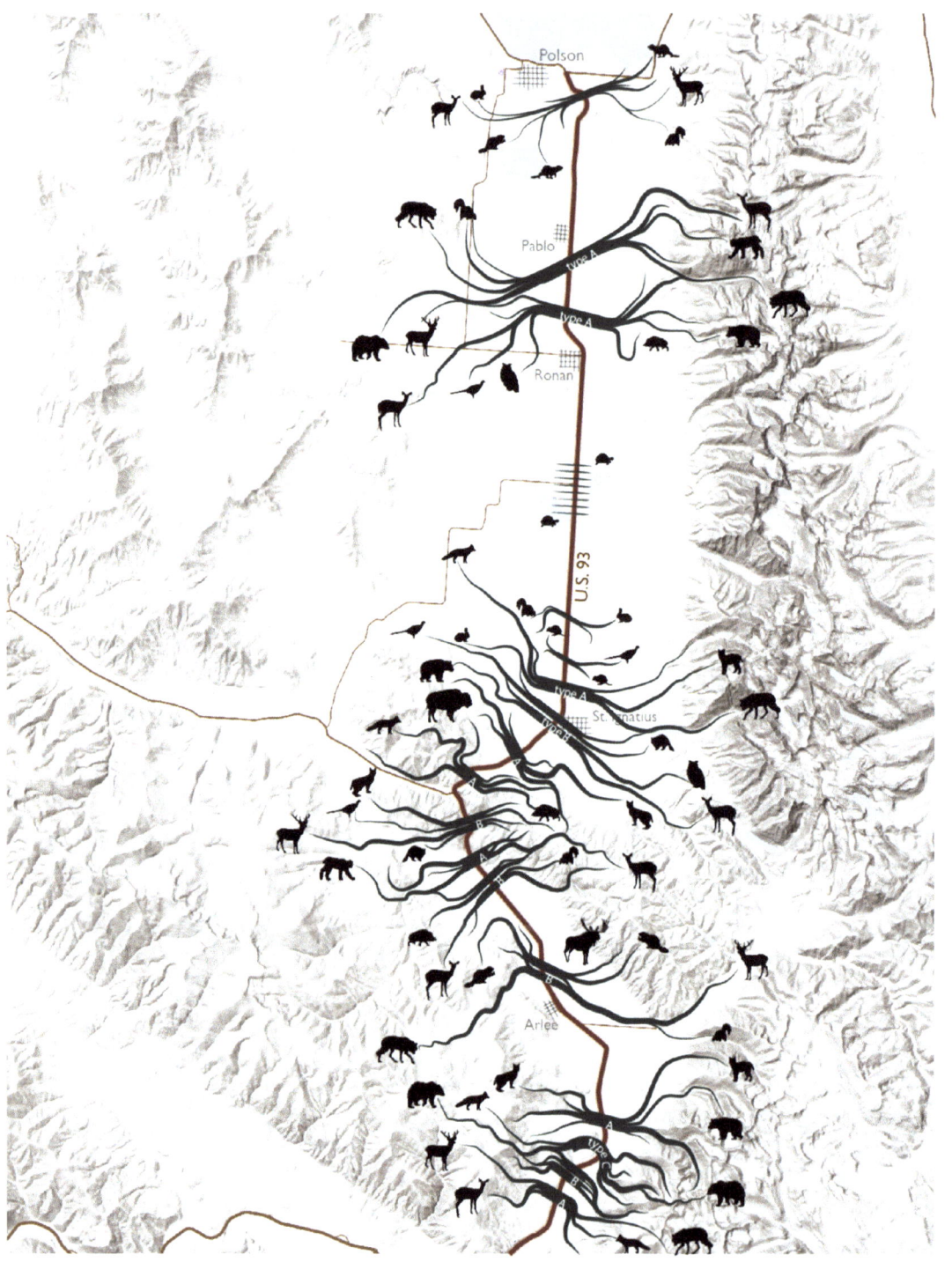

5.4 Wildlife migration and crossing location map.

The final MoU included 56 wildlife crossings. For the next three years, we wrestled with the engineers and tribal staff to understand the highway setting and peoples, its unique qualities and its potential for bringing communities together.[11] We designed prototypes of the crossings to fit the large mammals migrating between the Mission and Bitterroot Mountains. Crossings were located at stream crossings and natural breaks in the topography based on roadkill data and the feasibility of fitting a box culvert underneath a highway. Most importantly, we assembled engineers, wildlife biologists and tribal staff to visit each potential crossing location to discuss its siting, parameters and size in the field. This immersive decision-making became determinative of wildlife crossing decision-making for the rest of the project. Tribal staff used the site analysis report generated from these field trips throughout the design process.

Eventually, Julie, one of the originators of the wildlife crossing idea, moved to another firm. Jones & Jones was not asked to complete the final construction documents; MDT divided the highway project into eight segments and gave the work to local engineering and landscape architecture firms. The highway was constructed and named the People's Highway. It won awards for its incorporation of wildlife crossings and landscape features. Our firm's name did not show up on the awards. The wildlife crossings have helped several wildlife biologists do pioneering research, get their PhD, and receive credit for original research on wildlife migration. Implementation of the crossings have reduced the number of accidents on the highway, honoring the ecology, history and culture of the place.

I tell this professionally complicated tale not to denigrate the role of the engineers, many of whom worked diligently to solve the puzzle of incorporating wildlife crossings into a highway, but to illustrate that relations (positive and negative) define landscape dialogue. An understanding of relations and their qualities precedes project success. This is true for spatial relations of the existing landscape, in this case who drives, where do they go, where are the accidents and who are the players. This is true for "hidden" relationships such as people and wildlife. It is true of the design/client team itself. Who wants what? How do they see the world? Who else needs to be involved?

In places in North America (and certainly the rest of the world), the notion of the individual as an independent decision-maker is antagonistic to ideas of what it means to be in the landscape. In the Western Apache culture, "land" and "mind" are the same word, both a repository of memory and an

5.5 Types of wildlife crossings in the landscape (top left, small box culvert, Kootenay National Park, Parks Canada; top right, arch culvert near Polson, Montana, Jones & Jones; bottom right, wildlife overpass near Arlee, Montana Department of Transportation; bottom left, bridge over Finley Creek).

active force for meaning.[12] Keith Basso, an ethnographer who partnered with Western Apache elders to map places names in eastern Arizona, documented how place names informed Apache history, landscape and social relations.[13] In one example, the name Yúdahá Kaikaiyé (Navajos Are Coming!) indicates the draw where four Apache families sought refuge when a young woman alertly noticed the whinnying of an unfamiliar horse. When Apaches use a place name, they invoke the words of their ancestors which not only provide meaning to the place but relate present to past peoples. Place names are the connective tissue of Western Apache society, linking families and spaces and times. They have a different, more spatial conception of history, and thus identity. "What matters most to Apaches is where events occurred, not when, and what they serve to reveal about the development and character of social life."[14] The connection between place relations and social relations is indistinguishable.

From the point of view of Western Apaches, Anglo-American history does not relate to place. It is "removed from the contexts of daily social life (reading, Apaches have noticed, is an isolating activity), it seems unconnected to daily affairs and concerns."[15] It is "unspoken and unanimated," without relation to the present life of people. Landscape relations are communicated through active and ongoing dialogue, not reading a text. As Basso argues, Apache ancestors are not absent from place. Place-making, the buzzword of civic architecture, is a social process of comparing ideas, pursuing implications and harkening the words of ancestors which offer practical and moral guidance.

POWER GEOMETRIES

If landscape dialogue is relational, then designers must acknowledge these relations and their unevenness. Rarely do relations exist – between people, between the landscape and a community – based on equal reciprocity. Each suite of relations is contorted by power, an imbalance in relations whereas one group/entity/person influences another. Power manifests in close relationships, such as when a parent tells a child to clean her room, or in distant relationships, such as when a company advertises a soft drink that young people buy. "In reality power means relations, a more-or-less organized, hierarchical, coordinated cluster of relations."[16] While there are close power relations in the landscape, such as the Montana motor home driving slowly with traffic backing up behind them, this discussion will focus on distant relations as broader and more numerous distributions of power.

Max Weber, the influential German sociologist, described power as domination, the removal of choice from people who previously experienced freedom – the ability to get something done by someone who potentially could resist.[17] It is a bounding, a constraint. One person/group/entity removes choices available to another. Spatially, one can consider traditional forms of power to be like a monarchy, a central point from which power radiates outward. As such, traditional power can be diagrammed.

Yet power is not solely a transaction between individuals. It works between many groups and people and space and infrastructure. Michel Foucault, in his classic historic inquiry, describes power as immanent, as diffuse, remote, everywhere.[18] But also specific, a product of history, particular to a people. "Power functions, then, through people 'working' on their own conduct, fashioning themselves in ways that reflect their acceptance of a particular norm which, perhaps unwittingly, makes them subject to its control."[19] This is not the power of a monarch disseminating edicts from a throne, but power of the self-identification of his subjects as subject, who agree to not only the edict, but the monarchical structure which issues the edict. Foucault's power is not in the hands of the ruler, but in everyone who allows themselves to be ruled. This post-modern kind of power is more diffuse and, thus, more difficult to diagram.

The two models of power – central and diffuse – are not necessarily incompatible. It takes both kinds to maintain a system of uneven relationships, as Foucault acknowledged.[20] For the designer, the more immediate task is to identify power differentials in relation to the landscape. To do this, we first acknowledge power as spatial and aesthetic.

Power as spatial

Power is realized in space. In the landscape, Foucault's diffuse model of power misses the particularity of space and its differences. In a lucid account of power's relation to space, John Allen emphasizes this particularity, the way that power in space can take the form of domination, authority or seduction (among others).[21] Power is as much a collaborative exercise or agreement as it is an instrumental act to control others. Power is more relational, or associational, than instrumental.

Domination of space means choices are removed. Space is produced for a singular purpose, e.g. a U.S. highway for cars driving fast. The homogeneity of use requires the power of the State (Weber) *and* the agreed upon culture of driving (Foucault) to perpetuate its design and maintenance. For U.S. 93, the disruption of State power reliant on the engineer's worldview only happened under threat of legal action in the context of past treaties with a sovereign nation (the Confederated Salish and Kootenai Tribes). Even in that process, to suggest a different use of this linear space would have seemed absurd and was never considered.

The range of choices of what a highway could be was enlarged to include wildlife, but not enlarged to question whether driving at high speeds, with its concomitant threats to the climate and the people, was beneficial.

If power is ultimately spatial, then it can be described through geometric relations. Power flows through the landscape in the taken-for-granted infrastructures of the city. Infrastructure, political groups, resident relations can be diagrammed. "People are placed by power, but they experience it at firsthand through the rhythms and relations of particular places, not as some pre-packaged force from afar and not as a ubiquitous presence."[22]

Doreen Massey, the British geographer, describes the relations and directions of power as "power geometries."[23] She uses power to explicate globalization – the shifting spatial and economic relationships that extend across cities and countries.[24] Globalization affects people differently, some it empowers, some it constrains. These geometries are not static, positional outcomes, but dynamic and shifting processes. Close relations stretch and become distant. As goods and ideas flow faster and faster across space,

5.6 Power geometries: the weight of freeway concrete over people's (temporary) homes in Oakland, California.

it is increasingly apparent that these spaces are not a-temporal, that is static or a collection of objects. Rather, space is a process; it is dynamic, nodes of "overlapping trajectories." People use power to "hold the world still in order to look at in cross-section."[25] But the world is a process of becoming, meaning its relations also grow and decline. Power geometries describe hierarchies, people "above" and "below," and the amount of power between people and objects in space.

The forces of globalization can be seen in the Flathead Indian Reservation as the flow of goods along the highway from Missoula to Flathead Lake, as the profusion of scenic images of Glacier National Park that inspires tourists to drive their motor homes up U.S. 93 and even in the distinction between local and global, native and non-native, that is applied to plants, animals and people. Globalization compresses time and space through the faster and faster exchange of goods.[26] Long distances become less relevant at high speeds. This extends power to outside a community, to the layers of hierarchical organizations existing as part of the State, national and global governance.

While power in relations is often discussed as a negative thing, as oppression in the case of Weber and Foucault, it can be a positive force. Power moves things. Transforms people. Shapes the world. Power can be generous, as in a redistribution of goods for justice, an opening up of space for all. Power can be collaborative. Communicative. Negotiated. Engineers may have power bestowed by the people, by the State, but if they embrace alternative spaces and flows, i.e. the movement of wildlife beneath and above a highway, then power transforms previously uneven relations to more equitable spaces. Power then shifts from one group to another (and often back).

> The negotiations among those actively seeking to create an alternative set of spaces, to use them in ways that are less confrontational and more adaptive, may serve to dilute homogeneous spaces, but the intent may be to carve out a positive-sum scenario where a more beneficial use of space for all is envisaged.[27]

Power as aesthetic

Power is not just centered in those-who-make-decisions; it permeates how a space should be, what it should look like. It is the assumption that certain things happen in certain spaces and other things do not. It is the recognition that certain people go to certain spaces and others are out of place ... It is an aesthetic issue. In *Geographies of Power*, Allen puts it this way:

> The entangled nature of people's lives, places ... take their shape through dominant or controlling rhythms that seek to suppress the routine traces of others. Exclusion in this context has less to do with closed doors and high walls, and rather more to do with spaces constructed by dominant groups in their own likeness – through a series of rituals and gestures, moods and attachments, as well as accumulated styles and meanings. The composition of space, the partition and layout of particular uses, also serves as both a resource and the means through which power is exercised.[28]

The U.S. 93 landscape as produced is a visual pattern arranged to provide a ribbon of asphalt within, through and along it. This pattern has become prolific in the Western landscape – the way things should be, the way things are. Jacques Ranciére describes the "distribution of the sensible" as a narrative or visual regime, an appropriate way of describing the world, one which makes it difficult to describe the world in an alternative manner.[29] In the U.S. 93 wildlife crossing project, the original power scheme based on the needs of a general (driving) public is visible and assumed. What is left out of that relationship is what is less visible: tribal residents who might live there and the wildlife crossing the highway. Drivers moving at 65 mph cannot see the wildlife of the Flathead Indian Reservation unless they hit a deer. MDT mounted motion-sensing cameras within the wildlife crossings, taking pictures of river otters, deer, elk, black and grizzly bear. The pictures make the wildlife visible (at least to those who visit the website), moving them from the invisible landscape to a place of power (however slight). The tribes, scientists and MDT maintenance personnel use the images to measure and promote the highway's success.

Much of the natural and social sciences is learning the process of noticing the invisible, whether the molecular, the below ground, or the silent majority. Landscape relations are often invisible, only emerging from behind the veil, during conflict ... an animal gets hit by a car, a tribe sues the State. I explore the process of making the invisible visible through criticism and dissent in Chapters 6 and 7.

5.7 Matrix of wildlife using underpasses – from top left, clockwise: wild turkey, marmot, white tail deer, river otters, bobcat and kit, black bear, badger, mountain lion. Grizzly bear in center.

Landscape dialogue changes as power shifts. The dialogue a designer has with a client is different than a dialogue with a sub-consultant. Even the nature of "client" changes in relation to power and dialogue. In many cases, the landscape is the client. In the U.S. 93 project, some would say the wildlife was our client. Re-shifting power geometries, in this case, treating the tribes and the landscape as our "client," meant less prioritization given to maintaining our legal client, the engineers at the lead design firm and MDT. This re-shifting of "client" ultimately led to a successful project in terms of innovation and landscape health, but also led to our not being able to see the project through construction documents and implementation.

By assessing power geometries in the landscape, as part of landscape dialogue, the designer can better envision where the potential for change may be, and where it may be necessary to change underlying structure before true change can happen.

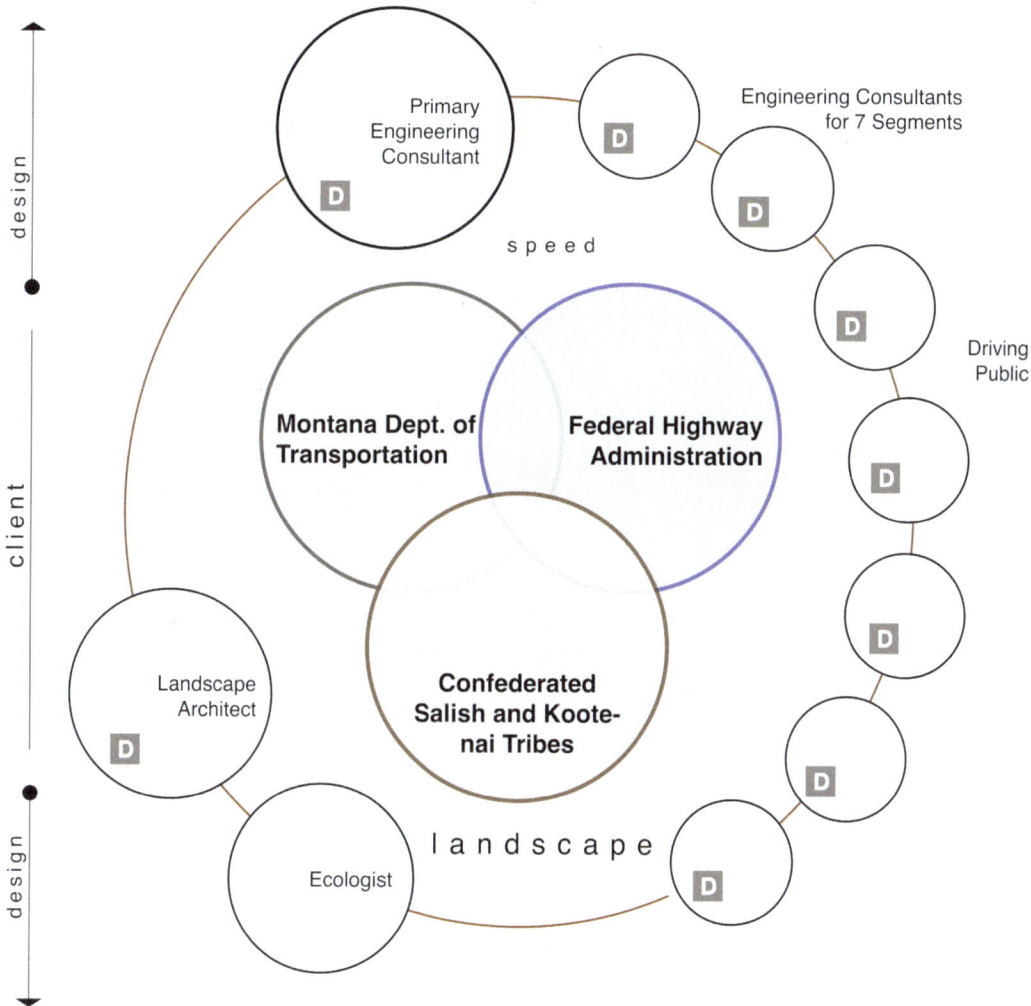

5.8 Design team relation diagram after MoU signed by the three main entities with an expanded client, including the landscape and wildlife.

PRAXIS

The designer engages with relations and power geometries in order to delve into assumed and neglected areas of design. I list some steps, not to suggest that the process of relations assessment is linear, but to move from general relations (i.e. who is the client?) to specific relations which require more detail. Praxis in relation to power should take place in the context of a real project, although historical projects can also be evaluated using the same framework.

An evaluative framework for assessing power relations includes:

1. Client definition – shifting "ownership" from a property focus to a landscape focus.
2. Visibility assessment – what things need to be revealed?
3. Diagram power geometries – evaluating the relations of visible and invisible actors in the landscape.

1. Define the client and its (their) relations

In design projects, clients range from a single entity, like a homeowner, to a complex network of entities like a state government. The contractual agreement negotiated with the client will influence how the project proceeds. At the beginning of a project, it is important to expand the definition of client. Who is this project for? If a single homeowner, then include the homeowner's grandchildren who frequent the backyard on weekends or the songbirds that frequent the neighborhood so the homeowner can set out a feeder. This client-defining step can wrestle with moral questions of environment and community. What is the nature of private property? Does it exist solely for the benefit of the property owner or does the broader community have a say? If the legal client would like to place an aesthetically inappropriate pink flamingo in the front yard, does the landscape architect have an obligation to the neighbors to discourage this? That is, is the client in this case the homeowner *and* the neighborhood? A designer focusing on the legal client may not account for power geometries. Neighbors can wield power in unexpected ways and make life miserable for pink-flamingo-loving homeowners.

In public projects, the client as a multi-headed hydra can become unwieldy. A design project may be solicited by a government agency with knowledgeable staff acting as the point of contact and the agency itself the signer of the design contract. Yet the people in the government agency work for the broader public. They abide by statutes and codes written for the public benefit. The U.S. National Park Service (NPS), for instance, operates under the Organic Act which famously describes the agency as stewarding the landscape of the parks to "leave them unimpaired for the enjoyment of future generations."[30] In an NPS project, the designer would have a legal contract with staff of the NPS but their client would be the public and "future generations," nebulous terms describing all people irrespective of time. The Organic Act gives power to people who are not born yet, who currently have no voice.

If U.S. 93 had been a traditional highway design project, wildlife would still have been part of any Environmental Impact Report for the highway. A wildlife biologist would have been dispatched to list the number of species in the area, to assess the impacts to wildlife from adding a couple of lanes and the assessment would have ended with a few recommendations to mitigate damages caused by a wider highway. With alternative clients, the purpose and process of the project changes. If the client is the landscape itself (and, yes, all the people driving within it) then the approach broadens even more to include myriad relations revolving around landscape health, wildlife movement and cultural practices.

2. Assess the visibility of relations

In the assessment of relations, the designer first inventories and evaluates the visible connections between landscape and people, people and groups and between landscape processes. In a relationship between people who visit an urban space and people who maintain the space, for instance, a number of questions can be asked. What is the visitor's interaction with and opinion of those who maintain the space? Is the maintenance effective? And if not, does the lack of maintenance affect the number and quality of visits? By answering questions related to maintenance, the designer can situate both the people involved in the landscape as well as the process of maintenance in the context of power geometry. In some urban open space, maintenance personnel may dominate decision-making, ensuring durable landscape elements and only specific uses, at the expense of creative expression in design. In other urban open spaces, visitors may disparage maintenance personnel because of the color of their skin or their lower-class status. The relationships to be assessed can be as simple as that between visitors and maintenance personnel or expand to include types of visitors and their relation to landscape elements: pavement, production of waste, deciduous and evergreen trees and types of soil.

Or it can be as simple as walking on pavement in high heels. In Occidental Square in Seattle, Washington, pedestrians walked across cobbles salvaged from historic buildings torn down in the 1970s. The pavers gave the plaza a rich texture and depth. They also made it difficult to walk in high heels. Over the years, people experiencing homelessness frequented the plaza, businesses complained and the future buildings to activate the space on its eastern edge never materialized. The square became a flash point of local politics and shifting power geometries.

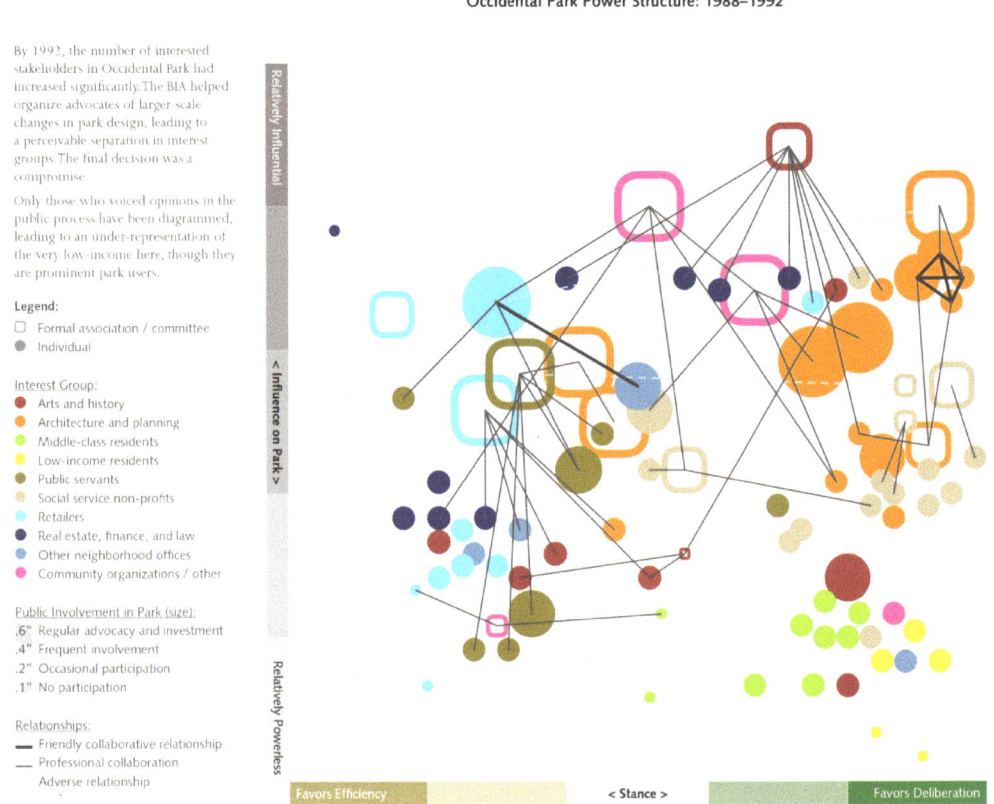

Occidental Park Power Structure: 1988–1992

By 1992, the number of interested stakeholders in Occidental Park had increased significantly. The BIA helped organize advocates of larger-scale changes in park design, leading to a perceivable separation in interest groups. The final decision was a compromise.

Only those who voiced opinions in the public process have been diagrammed, leading to an under-representation of the very low-income here, though they are prominent park users.

Legend:

☐ Formal association / committee
● Individual

Interest Group:
● Arts and history
● Architecture and planning
● Middle-class residents
● Low-income residents
● Public servants
● Social service non-profits
● Retailers
● Real estate, finance, and law
● Other neighborhood offices
● Community organizations / other

Public Involvement in Park (size):
.6" Regular advocacy and investment
.4" Frequent involvement
.2" Occasional participation
.1" No participation

Relationships:
── Friendly collaborative relationship
── Professional collaboration
    Adverse relationship

Relatively Influential

< Influence on Park >

Relatively Powerless

Favors Efficiency   < Stance >   Favors Deliberation

5.9 Occidental Square relational diagram showing power geometries of 1988–1992 based on influence and public involvement. In this time period, community and art organizations, as well as architects and planners were most influential and collaborative.

Boting Zhang, in a thesis completed at the University of Washington, diagrammed the power dynamics that existed in 1980, 1990 and 2006.[31] In Zhang's treatment, the reader can see relations, values and power described as "influence." One can also see a shift over time from power in the hands of non-profit groups to power in the hands of retailers and civic government. The shift resulted in a remodeled square with smooth concrete pavers, less fixed seating and the removal of structures (which provided shelter from the rain for the square's indigents).

Zhang's analysis of Occidental Square did not include other relations more difficult to entangle due to their lack of visibility. People experiencing homelessness, for instance, do not show up in the relational diagram due to their "invisibility" in terms of decision-making (although local organizations did advocate for them). It did not occur to the city they might have opinions on the plaza design, although their visible presence

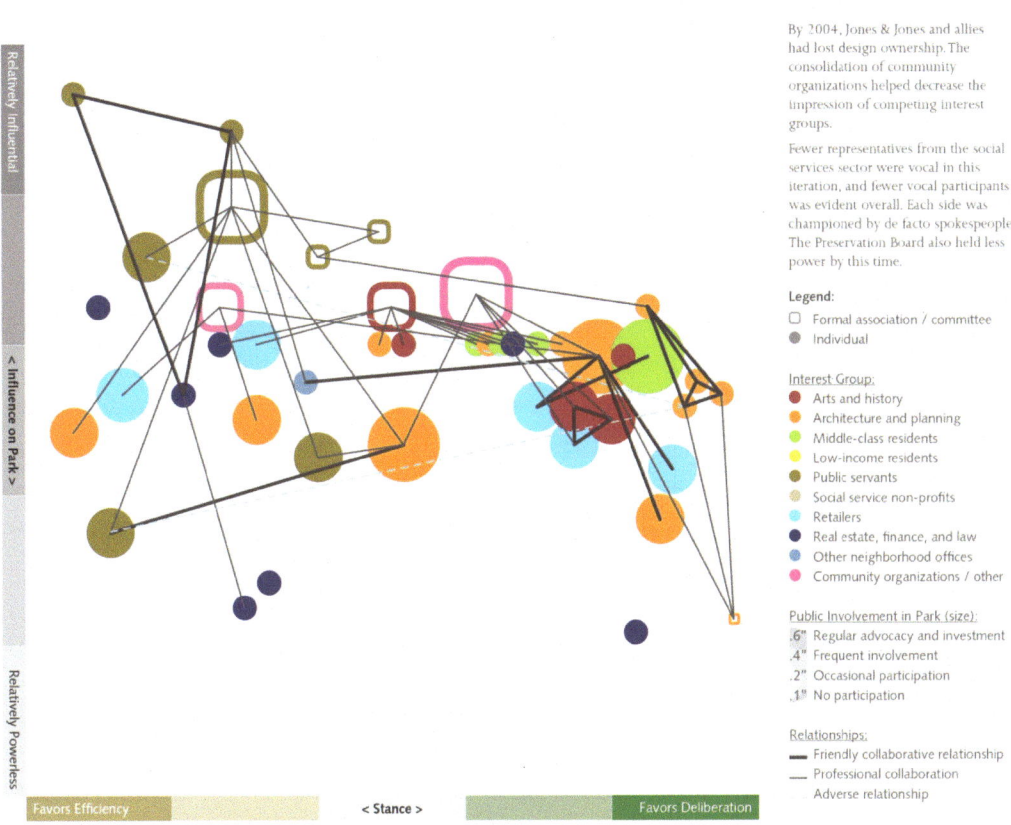

Occidental Park Power Structure: 2004–2006

By 2004, Jones & Jones and allies had lost design ownership. The consolidation of community organizations helped decrease the impression of competing interest groups.

Fewer representatives from the social services sector were vocal in this iteration, and fewer vocal participants was evident overall. Each side was championed by de facto spokespeople. The Preservation Board also held less power by this time.

Legend:

◻ Formal association / committee
● Individual

Interest Group:
● Arts and history
● Architecture and planning
● Middle-class residents
● Low-income residents
● Public servants
● Social service non-profits
● Retailers
● Real estate, finance, and law
● Other neighborhood offices
● Community organizations / other

Public Involvement in Park (size):
.6" Regular advocacy and investment
.4" Frequent involvement
.2" Occasional participation
.1" No participation

Relationships:
━━ Friendly collaborative relationship
── Professional collaboration
Adverse relationship

Relatively Influential

< Influence on Park >

Relatively Powerless

Favors Efficiency   < Stance >   Favors Deliberation

5.10 Occidental Square relational diagram showing power geometries of 2004–2006 based on influence and public involvement. In this time period, power has shifted to local retailers, city government and more top-down decision making.

was one of the instigating events for the plaza remodel. A key component in assessing invisible relations is determining who is not at the table who should be.

Non-design disciplines can help make invisible relations visible. Science finds links between air quality and children's health or nutrient cycling in urban soils. Sociology examines the behavior of street youth in public space. Each exploration into social and ecological context reveals new insight into the landscape. History, at least the historic approach of the intensely local such as practiced by the Western Apache, yields stories of relations, now obscured, but significant to meaningful places. Past trauma writ in the landscape – pollution, violence, extraction – heals slowly. Broken relationships must first be acknowledged before healing/ design can begin.

3. Construct a relational diagram

The last "step" in a relational evaluation based on power is to construct a relational diagram of power. If power is neither diffuse (as in Foucault) nor concentrated (as in Weber), but a shifting web of relations, then how does one graphically represent basic relations in that network? List primary actors in the narrative of the landscape. Include the landscape itself, potentially dividing the landscape into different elements or areas if necessary (i.e. wildlife, water, residents). Ask questions of the separate actors ... who decides the shape of the space? Which landscape elements are important to the community? How do these elements relate to each other? Then, assemble a diagram using a relational symbology to represent different elements of power. This relational diagram can be used in negotiations with the potential configuration of the space, or it may remain in the background, guiding evaluation from behind the scenes. Either way, assessment of power and relations over time and space situates landscape dialogue.

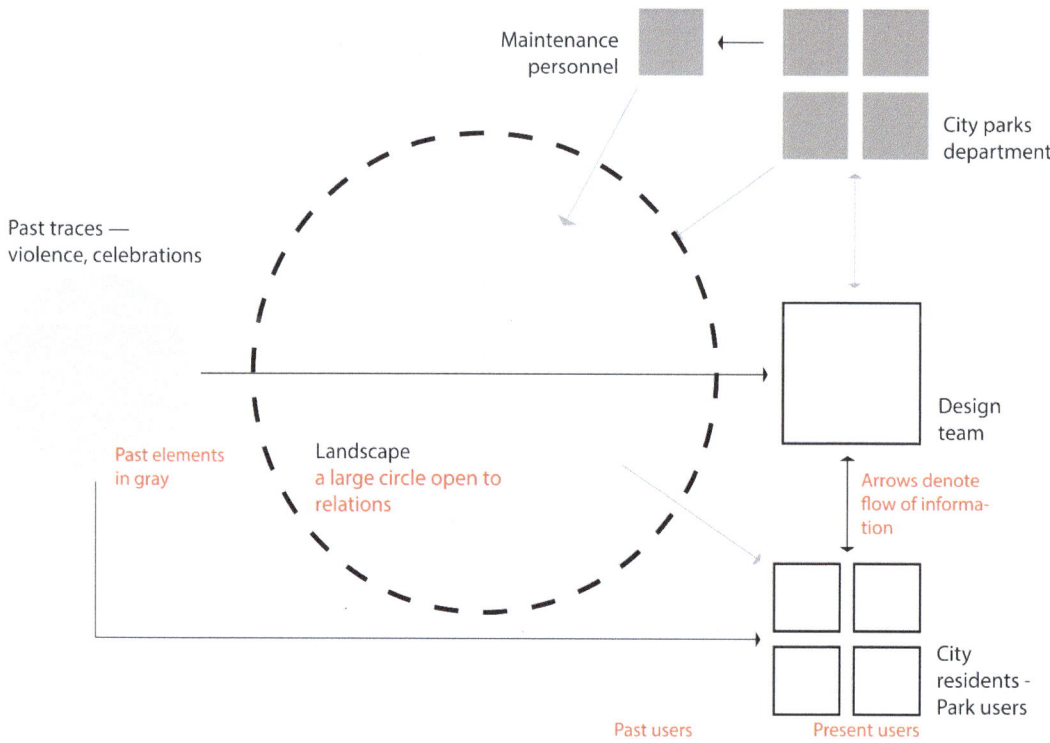

5.11 A sample symbology for a relational diagram.

NOTES

1. Carlo Rovelli, "The Big Idea: Why Relationships Are the Key to Existence," *The Guardian*, September 5, 2022, sec. Books, https://www.theguardian.com/books/2022/sep/05/the-big-idea-why-relationships-are-the-key-to-existence.

2. Confederated Salish and Kootenai Tribes, *Cultural Resource Overview for U.S. Fish and Wildlife Service Western Management Properties* (Pablo, MT: CSKT, 2000).

3. Landscape Architecture Foundation, "Guide to Evaluate Performance: Landscape Performance Series," Landscape Performance Series, November 6, 2014, https://www.landscapeperformance.org/guide-to-evaluate-performance.

4. Rhonda Swaney, "Letter to Joel Marshik Described as 'Tribes' Preferred Alternative'," in *Final Environmental Impact Statement and Section 4(f) Evaluation for U.S. 93 Evaro to Polson, Montana* (Helena, MT: MDT and FHWA, 1996).

5. There is a great deal of literature on speed and speed limits, some of which specifically focuses on Montana. Engineering articles focus on instrumental rationality, offering valuable insight into human behavior, but not accounting for the social context or even the design of the road (for instance, Gayah et al. 2018). Other articles on speed address some of these social and legal contextual considerations (i.e. Robert E. King and Cass R. Sunstein, "Doing without Speed Limits," *Boston University Law Review* 79 (1999): 155–171).

6. Marcia Pablo, "Preservation as Perpetuation," *The American Indian Quarterly* 25, no. 1 (2001): 18–20.

7. Dale Becker, *Highways and Movement of Wildlife: Improving Habitat Connections and Wildlife Passageways Across Highway Corridors, Proceedings of the Florida Department of Transportation/Federal Highway Administration Transportation-Related Wildlife Mortality Seminar* (Washington, DC: Federal Highway Administration, 1996).

8. Dale Becker, W. Camel, C. Parker, C. Scott and I. Luters, "Wildlife Crossings within the Cultural Landscape of the Flathead Indian Reservation: U.S. 93 from Evaro to Polson, Montana" (International Conference on Ecology and Transportation, Seattle, Washington, 2011).

9. Montana Historical Society Press, *Coyote Stories of the Montana Salish Indians*, 1st edition (Helena, MT: Montana Historical Society Press, 1999).

10. Becker, *Highways and Movement of Wildlife*.

11. Becker et al., "Wildlife Crossings within the Cultural Landscape of the Flathead Indian Reservation."

12. Keith H. Basso, *Wisdom Sits in Places: Landscape and Language Among the Western Apache*, 1st edition (Albuquerque, NM: University of New Mexico Press, 1996).

13. Basso, *Wisdom Sits in Places.*

14. Basso, *Wisdom Sits in Places*, 31.

15. Basso, *Wisdom Sits in Places*, 33.

16. Michel Foucault, "The Confession of the Flesh," in *Power/Knowledge: Selected Interviews and Other Writings*, (New York, NY: Vintage Publications, 1977): 194–228.

17. Max Weber, *Economy and Society: An Outline of Interpretive Sociology* (Berkeley, CA: University of California Press, 1978).

18. Michel Foucault, *Discipline & Punish: The Birth of the Prison*, translated by Alan Sheridan, 2nd edition (New York, NY: Vintage Books, 1995).

19. John Allen, *Lost Geographies of Power*, 1st edition (Malden, MA: Wiley-Blackwell, 2003), 76.

20. Michel Foucault, *Language, Counter-Memory, Practice: Selected Essays and Interviews*, 1st edition (Ithaca, NY: Cornell University Press, 1980).

21. Allen, *Lost Geographies of Power.*

22. Allen, *Lost Geographies of Power.*

23. Doreen Massey, "A Global Sense of Place," *Marxism Today* 38 (1991): 24–29.

24. Massey, "A Global Sense of Place."

25. Doreen B. Massey, *For Space*, 1st edition (Los Angeles, CA: SAGE, 2005), 36.

26. David Harvey, "Time-Space Compression and the Postmodern Condition," *Modernity: Critical Concepts* 4 (1999): 98–118.

27. Allen, *Lost Geographies of Power*, 187.

28. Allen, *Lost Geographies of Power*, 11.

29. Jacques Rancière, *Dissensus: On Politics and Aesthetics*, translated by Steven Corcoran, Reprint edition (London: Bloomsbury Academic, 2015).

30. United States Congress, "National Park Service Organic Act," 16 U.S.C. 1,2,3 and 4 § (1916).

31. Boting Zhang, "Occidental Square: How Community Power Structure Impacts a Park in Seattle's Pioneer Square Neighborhood" (Master's Thesis, Seattle, Washington, University of Washington, 2011).

CHAPTER 6

Landscape critique: critical engagement and violence

In 1598, on the Catholic day of Ascension, long before the United States or even the 13 colonies existed, Juan de Oñate stood on the south bank of the Rio Grande, assembled his exploration party for a prayer and claimed all lands north of the river for New Spain. He and his party of Spanish, mestizo and black colonists proceeded up the Rio Grande to arrive at Ohkay Owingeh near present-day Española. Oñate established and became the first governor of the Province of Santa Fe de Nuevo México. Today, Oñate is honored as the "George Washington" of the Hispanic peoples of the American Southwest.[1] His name adorns schools, parks and streets in New Mexico and Texas. Oñate, for many, represents the beginning of Hispanic culture, Catholicism and the Spanish language in the region.

Four hundred years later, during the Cuartocentennial celebrating Juan de Oñate and the establishment of New Mexico, a journalist for the *Albuquerque Journal* received a picture of a bronze boot in the mail. Along with the boot, a note stating:

> We invite you to visit the Oñate Distortion Museum and Visitor Center located eight miles north of Española. We took the liberty of removing Oñate's right foot on behalf of our brothers and sisters of Acoma Pueblo. This was done in commemoration of his 400th year anniversary acknowledging his unasked for exploration of our land. [Signed: Friends of Acoma]

Sure enough, when a call was made to the Oñate Museum and Visitor Center, staff confirmed that the heroic figure of Oñate on horseback was missing a right foot. Why was the bronze foot removed? And what does

DOI: 10.4324/9781003158943-7

6.1 Juan de Oñate sculpture at the Oñate Museum and Visitor Center near Espanola, New Mexico.

this have to do with the "brothers and sisters of the Acoma Pueblo?" Somehow, events amidst the conquest of the region influenced present-day relations between Hispanic peoples and the Acoma peoples, as well as the landscape itself.

To understand, we must return to 1598 and the beginning of Spanish colonization. After setting up a provincial government, Oñate's nephew led a party west of the Rio Grande to search for passage to the Pacific Ocean. Bereft of food and supplies, they encountered Acoma Indians (*ʔáák'uumé*) at the base of their pueblo and asked for corn and provisions. When the Pueblos did not provide the corn after a few days, 15 Spaniards climbed the steep pathway up the 300-foot cliffs and dispersed through the village. One took two turkeys (sacred to their culture) and two more harmed Acoma women, before the Pueblos attacked and killed them. In response, Oñate sent his brother to lead an expedition of 70 men with two cannons to revenge their deaths. The resulting battle/massacre resulted in the killing of 800 Pueblos and one Spaniard. Oñate's brother brought

6.2 Sky City of Acoma Pueblo, one of the longest continuously inhabited towns in the United States.

500 prisoners back to Ohkay Owingeh, where they stood trial. All men older than 25 years were found guilty, sentenced to 20 years of slavery and their right foot was cut off.

It has been over 400 years since this massacre took place. It is a long time. And yet, violence persists. The Acoma poet Simon J. Ortiz puts it this way:

> Memory we cross and cross again. Treks, trauma, and on.
> We do know what time is. It is loss and gain. A lingering.[2]

Friends of Acoma sent a picture of Oñate's foot to speak a story of violence from long ago in the hopes of remembering a culture that was irrevocably changed by Spanish colonization. To quote from their letter to the *Santa Fe Reporter*:

> New Mexico was poised for a grand celebration of the Cuarto Centenario and we could not let that happen without voicing our existence. Outside of "Indian art" and "gaming," we have become an invisible people, even to ourselves. Our Hispanic brothers have forgotten on whose land they dwell … Our actions were to redirect the thinking of those who have forgotten us.[3]

The celebration of an individual such as Oñate by one culture does not always erase or diminish the stories of another culture, but in this instance, it has.

Landscape dialogue is a conversation with the past landscape as well as the present. As designers, it is important to understand past events in a landscape, particularly from cultures other than your own. It requires attentive listening. Cutting off a foot on a statue is a critique of the landscape, a critique of what gets memorialized in that landscape. Often, critiques are not this dramatic and direct. As part of the listening process, designers dig into the history of a place to uncover memories of conflict and peace. Where can we find the stories of a place in more banal circumstances and places? We talk to people who live there and ask them for stories of family history. We talk to people who have been displaced … who used to live here? We measure their displacement, both a physical measure of distance and an emotional measure of their disconnection from the place. These "measurements" are not in search of an absolute value but an immersion into displacement: what it means for the displaced, what it takes to stay connected with place, what stories are told about and told in spite of the displacement.

In the United States, this can be difficult. Native Americans, the most frequently displaced people groups, are reticent to share their experiences with the descendants of their oppressors. In this case, the designer can still find stories in book compilations and local libraries. For U.S. 93 (see Chapter 5), there was a remarkable wealth of information online and in the University of Montana library in Missoula. So much so, that the tribe's wildlife biologist had the cultural resource officers review an article I wrote to make sure I wasn't revealing anything the tribe didn't want revealed.[4]

The event at the Acoma Pueblo spans centuries and large swaths of New Mexican territory linking Spanish colonization (and later Anglo colonization) with the spatial characteristics of the Acoma Pueblo and northern New Mexico. The isolation of the Pueblo, its impenetrable cliffs amidst the high desert (Figure 6.2), harbor memories – for some, generations of life, family and culture, for others a romanticized "empty" space ripe for conquest. Through the story of Oñate's brutality, this particular Pueblo and its landscape connects to the provincial capital, Ohkay Owingeh, as well as the Oñate Museum and Visitor Center, the site of the statue whose foot was cut off. A critique of the site and the sculpture suggests there is no site as an isolated phenomenon; a celebration of historic men is not an individual, neutral story of "founding" a place. The past influences the local, present-day cultures and peoples of New Mexico.

> The case study [of Oñate's foot] marks a discursive turning point, for OMVC is a site of memory where commemoration is demonstrably shifting from the monumentalization of a historic event personified in a representative figure … to the memorialization of a process of cultural encounter and change acted out, in and tied to space.[5]

Therefore, which stories are told becomes part of the landscape and must be examined completely as part of landscape dialogue.

During the Cuatrocentennial celebrations, the city of Albuquerque decided to erect their own monument to Juan de Oñate in the center of Old Town (on a horse, of course). And here, things get messier. Until now, the story of Oñate's foot is a simple tale of dramatic oppression. But for the next few centuries after the massacre, the Acoma Pueblo would join forces with the Spanish military to beat back traditional enemies, like the Apache and Ute. Pueblo Indians intermarried with Spanish colonials (although Pueblo women were also subject to forced "marriages"). A blending of cultures for 400 years resulted in more complicated identities than the "tri-ethnic harmony" of present-day New Mexico promoted by civic authorities.[6] In response to issues of complex identity and the preservation of culture, it has become popular amongst Hispanics to research and portray the genealogy of their Spanish ancestors – to celebrate their Spanish identity, their historical-ness, despite the complexity of their familial backgrounds. In the face of American commercial culture, local Hispanics turn to Oñate who they see as the "father of Hispanic culture and their state."[7] This despite the fact that Spanish provincial leadership in the 1600s eventually retracted Oñate's governorship, recalled him to Mexico City and censured him for acts of cruelty.

The City of Albuquerque asked two artists (one Hispanic, one Anglo) to design a sculpture of Oñate. After protests from the Acoma and larger Pueblo and Indian activist community, they engaged a third, Nora Naranjo-Morse, a Tewa artist from Santa Clara Pueblo. At their first meeting, the two artists told Naranjo-Morse that she could design the base of their already-designed Oñate sculpture on a horse. The meetings deteriorated from there. Meanwhile, Oñate's bronze foot briefly returned during a protest in Albuquerque, held up by local activists at the site of the proposed sculpture.[8]

6.3 Reynaldo Sonny Rivera's *La Jornada* showing Oñate and settlers of New Mexico. A compromise.

6.4 View from the center of Nora Naranja-Morse's "Numbe Whageh" – Native American land art as a descent into the earth.

In the end, the protests and discussions culminated in a city council meeting in which supporters of the Acoma tribe, while outnumbering Oñate supporters, were unsuccessful; the city had devoted too much time to the statue project. They divided the art project in half. The Hispanic and Anglo artists designed and built a compromise sculpture of Oñate flanked by peaceful settlers to be placed on one side of the square. Nora Naranjo-Morse designed and built an earth sculpture out of desert materials, on the other side of the plaza. She set a descending spiral into the landscape so visitors could walk into its center as surrounding buildings disappear. In the center of the spiral, even Oñate and his fellow colonizers cannot be seen. The blending of cultures, the tri-ethnic harmony, the mestizo who contains in her body dual cultures and genes, could not be replicated, could not be constructed in a unified art form. It was too contentious.

How does one evaluate past (and present) violence in the landscape? Should it be commemorated? Or obliterated and forgotten? Or should the stories be told in multiple ways, as happened in Albuquerque's Old Town? All too often, design is complicit in retelling particular stories of the dominant culture's triumphs at the expense of other stories from other

6.5 Collage of Acoma Pueblo, Oñate's Espanola and Old Town sculpture's links with the past.

people. To extract the process from cultural complacency requires *critical thinking* – to look behind the veil of the dominant culture and uncover meaningful histories. The cutting of Oñate's foot is landscape critique, a criticism of the idea of a heroic great man (who, it turns out, was not so great) celebrated in the landscape of New Mexico. (In another sense, it is landscape dissent, a transgression of landscape norms, which I cover in the next chapter.)

CRITICAL THINKING

To think critically structures education, propels writing[9] and, most importantly for our purposes, drives dialogue.[10] What is critical thinking? Derived from the Latin word *crit*, critique means to judge, separate or decide. Primarily, critical thinking is judgement using reason. It is the skillful consideration of a landscape. Secondarily, thinkers of critical thinking have assumed a dispassionate consideration is required; that is, an observer at a distance. They describe criticism as "analysis and evaluation: analysis in terms of an explicitly stated framework, evaluation against explicitly stated standards."[11] This sounds sensible … how can you critique something without some sort of external measure with which to critique? Yet analysis removes the critical from critique; judgement becomes a self-perpetuating and narrow evaluation of success. Landscape critique should not remain the privilege of those who set "frameworks" or generate "standards." It is a process applied to the everyday landscape. Barr and Griffiths describe one of the challenges of an analytical "critique" by drawing from Arendt:

> What is normally intended by the notion of critical thinking (impartial, detached) fails when it comes to seeking to understand unprecedented events because such events bring to light the "ruin of our categories and standards of judgement". Such events demand, in her memorable phrase, "thinking without a banister". They demand explicitly judgmental storytelling.[12]

As we see in the case of Oñate's foot, an "impartial" or "detached" criticism fails to account for violence, situated as it is within a system that generate standards which arise from the protection of the very same system (of violence). Standards or frameworks may have little relationship with the local landscape; they affirm the detached as criteria without attending to the local events forming a place.

If critical thinking regarding the landscape is not the application of agreed-upon standards, what is it? Shirvani defines it as "both question and questioning the process of knowing the right question to ask … it is the ability to think reflectively in the process of making the landscape."[13] We first identify the need to question, instead of accept prior frameworks for site analysis. Why is so much energy and time spent on road design and construction at the expense of social gathering spaces? What flows exist outside of this site that influence the landscape and this project? Criticism needs to be broadened, not just with the questions asked, but with who is asking them. Reflexivity is the process of asking oneself questions regarding our place in the landscape. From which point of view, are we questioning? And which questions are we not asking, because of that position?

Landscape critique practices critical thinking as a subjective practice but a subjective practice with common elements. As bell hooks explains:

> When we engage in critical thinking [there] is an intensification of mindful awareness which heightens our capacity to live fully and well. When we make a commitment to become critical thinkers, we are already making a choice that places us in opposition to any system of education or culture that would have us be passive recipients to ways of knowing.[14]

Landscape critique then includes: (1) mindful awareness; (2) opposition to existing systems; and (3) active engagement. In addition to bell hooks' insights, I would add that landscape critique also (4) attends to the situational or spatial in the landscape. I examine critical thinking by expanding on these four characteristics.

Critique reflects on one's position

Landscape critique starts with mindful awareness. A reflexivity of thought considers potential connections and absences in our own life and our own discipline that might inform our critique. Part of reflective criticism is an understanding of *from where* we perceive the landscape in order to think critically. I delve into a designer's personal position and reflectivity in Chapter 8. Here, I want to emphasize where mindful awareness leads … the motion of people and the landscape. Our position in the landscape is always in relation to others; in the dialogic landscape, it is the others which shape our position. The challenge of reflexive critique is acknowledging both the separateness of ourselves from the other as well as the contiguous and intertwined nature of ourselves and the other … whether this other is another person or the landscape. Hannah Arendt, the twentieth-century philosopher, extended reflexivity to others by "training the imagination to go visiting."

> Critical thinking is possible only where the standpoints of all others are open to inspection. Hence, critical thinking, while still a solitary business, does not cut itself off from "all others." To be sure, it still goes on in isolation, but by the force of imagination it makes the others present and thus moves in a space that is potentially public, open to all sides; in other words, it adopts the position of Kant's world citizen. To think with an enlarged mentality means that one trains one's imagination to go visiting.[15]

It is communication with others that deliberately imagines oneself in another's position. If Emanuel Kant says the imagination is necessary to distance oneself from the observed, Arendt agrees but expands the imagination's role to then re-close the gap we experience between ourselves and others … and by extension between ourselves and another place. This includes the landscape. As she says, we need to be "strong enough to bridge the abyss of remoteness until we can see and understand those that are too far away as though they were our own affairs."[16] Judgement (or what I am calling dialogue) requires both a distancing of the imagination as a separate observer and an immersion in the place.

If we are to use our imagination to think critically about a place and its people, then the most difficult people and landscapes to imagine being in their place would be the people who are most different from ourselves. We are immediately faced with a problem: the educated designer, the architecture professor and the city planner, likely inhabit a different social sphere and often a different ethnicity than the disenfranchised.[17] It is crucial that designers place themselves in the shoes of the disenfranchised, for the production of space by dominant groups will be both visible and taken for granted. It is the invisible movements and productions of the disenfranchised we are concerned with in critical thinking. Critique requires stepping out of the dominant, accepted milieu and occupying a strange (to ourselves) position.

In discussing the marginalization of Jews, Arendt divides those who exist outside of dominant cultural milieus into three groups: parvenues, pariahs and conscious pariahs.[18] The parvenue assimilates as best she can into the dominant world, embracing the language and behavior of her oppressor. The pariah separates himself from the dominant culture, embracing fellow pariahs in the "warmth of persecuted peoples."[19] The conscious pariah, neither assimilates nor separates, but accepts difference in themselves and others to yield a critical understanding of the world. Arendt describes this state of being as "homelessness," which leads to a community of plurality.[20] Life in the margins. Then Arendt expands the idea of conscious pariah to include artists and writers, people who see the world differently (critically) – people who have trained their imaginations to go visiting. The writer becomes a conscious pariah as she tells stories that make the normal appear abnormal. The artist becomes a conscious pariah as he creates objects that challenge people to view the world differently. Designers become homeless … a community of life in the margins. The position of the conscious pariah is one from outside *and* inside – for if it was just outside, no one inside would be able to hear. And if it was just inside, nothing different would be said or heard. Critical reflectivity is an acknowledgement of our position and then re-positioning ourselves as conscious pariah.

parvenue

assimilation into the dominant culture

pariah

rejection of the dominant culture

conscious pariah

a critical approach to the dominant culture; an accep-tance of difference

6.6 Diagram of Arendt's parvenue, pariah and conscious pariah relating to the landscape.

Critique is oppositional

Landscape dialogue asks critical questions. Dialogue includes positive and negative comments, questions, affirmations and challenges, disagreements. The philosopher Mikail Bakhtin, who we met in Chapter 3, requires different points of view between two people to maintain dialogic relations acted out in space.

> I am conscious of myself and become myself only while revealing myself for another, through another, and with the help of another … Separation, dissociation, enclosure with the self is the main reason for the loss of one's self. The very being of man is the deepest communion … To be means to be for another, through the other, for oneself.[21]

According to Bakhtin, a complete understanding between two people (or two entities) would eliminate the need for dialogue, so there is always some Other that remains other. Dialogue is a permanent, ongoing need. The Santa Fe newspaper reporters receive a bronze foot in the mail. A story forms; questions revolve around the purpose of stealing the foot. Questions lead to discovering alternative stories related to people and place. These stories may rest in opposition to the viewpoints of other groups, but at the very least they oppose dominant ways of story-telling learned in school or heard at a city council meeting.

> In addition to codifying language and projecting new directions, criticism has … the duty to "exasperate, to increase the unease" of a discipline. This unease is frequently a function not of commenting on what was done, but on what was not done or said, on the silences within a project that bespeak much about situational or worldly meaning.[22]

The opposition, the Other found in dialogue, can be a person, a landscape or a larger system (i.e. real estate development). Harvey and other Marxist geographers define critical thinking as the process of revealing that which obscures (in their case, the hidden structures of an ideological and material landscape of uneven development).[23] In capitalism, the best worker is the one who cannot see their place in the work, but happily produces surplus value or profit for others, contributing to economic inequality. Thus, the first step in the struggle for equality is for the worker to realize his plight behind a veil of capitalist deceit. But to say that critical thinking is Marxist thinking is too narrow. Peeking behind the curtain to expose the empty arguments of power is central to most religions, post-modernism and the Wizard of Oz. The exposure of a deeper, but hidden structure to life assumes there is "more to the story." Part of that exposure, that peeking behind the curtain entails an acknowledgement of Power or powers working against the exposure. Critique must oppose this larger and "reasonable" system of land development. Landscape critique uses Arendt's story-telling – alternative histories like Oñate's foot – to explicate and contradict existing systems.

Critique is active and engaged

In landscape dialogue, the purpose of landscape critique is to let the landscape speak and then speak into the landscape (in a critical manner) through engagement and design. Outside of client standards and objectives from which to base one's landscape critique, what methods does the designer use to critique?

Critical inquiry is active. Critique happens in site selection, evaluation and the design process. It happens after design implementation. Intervention in a landscape – the placement of a statue, the digging of a trench – critiques past forms and spaces, present before the intervention. Critique is often thought of as promoting an architectural style (e.g. deconstructivism) by disparaging the assumptions of another style (e.g. modernism), but this can become passive, decoupling the landscape from its context and its people. Rather, landscape critique engages with the social, the personal ... the person standing in the street with cars streaming around or the squatters sleeping in an abandoned building.

Landscape dialogue incorporates design – the drawing of a positive vision, the speaking into a place and the speaking against inequities – in order to critique. The designer moves from language to form-making, which is itself a type of critique, to shape the world in a new way that "critiques" the old. Meyers uses the example of Richard Haag's Gas Works Park, questioning normative ideas of aesthetics and beauty of traditional parks.[24] Design uses existing and aspirational language (of landscape) to argue for another type of space.

6.7 Gas Works Park, Seattle, Washington. Design by Richard Haag.

Critique is situational

Landscape critique – of existing landscapes, of existing processes, of existing styles – is always situated in space. It matters where a place is and how it fits within a larger network. Meyer emphasizes both content and context in the process of design:

> Context refers to the meaning of the work's relationship to its immediate surroundings and to its cultural and environmental milieu. In brief, landscape architectural design as criticism should foreground the meaning of relationships among things, spaces, and systems in addition to the things, spaces, and systems themselves.[25]

The inclusion of context, the relations of "things spaces and systems," contributes to landscape critique. That is, context critiques project designs installed within site boundaries without attending to the larger spatial processes, whether high design or everyday vernacular landscapes.

By examining the surrounding elements and processes of a place, one can quickly find a site's limits, its narrowness.

Consider the site in Old Town Albuquerque that was selected for the planned statue of Oñate. As a site and just a site, its purpose was to present Oñate as an uncomplicated founder of Hispanic culture in the historic neighborhood of the largest city in New Mexico. Its situation pulls from a specific tradition of heroes on horses and pushes the larger context away. But think of the site as a landscape, as a spatial area connected through wagon trails, paths and highways to other parts of the region, as a temporal continuum that include moments such as Oñate's arrival in the region, but also his mismanagement of the territory and his eventual disgrace, as well as the Cuatrocentennial of his arrival and this suggests that the honoring of this person using a heroic gesture would be grossly inadequate. They are all part of the same story, an ongoing dialogue between Pueblo Indians, Hispanic communities, a specific place (Old Town) and northern New Mexico.

Without limits to the spatial and temporal context, how then does the designer select what to critique? To assess critically both content and context can overwhelm the designer. The site analyst must select significant landscape elements from the vast complexity of landscape processes, say soils, plants and existing visitor behavior. This selection, in itself, is a critical act. Choosing to examine taken-for-granted landscape elements, for instance, challenges the idea that these elements, as they currently exist, are assumed and inevitable aspects of the landscape. If the threat of a tsunami is what is significant, then one is saying other landscape elements – different ethnic groups, local economies or soil erosion – are less significant (to the landscape design). With limited attention and resources, designers must make a choice to focus on the critical aspects of a landscape, understanding in a profound way the history and potential of a place from this selective lens. As we practice landscape dialogue, how do we determine what is significant? I offer two potential ways of addressing this problem …

First, keep the landscape whole for as long as possible, the content and the context. This is a foundational principle of landscape dialogue. Do not separate the various layers of site information. Critique becomes a critique of the whole space as a landscape, as a system of processes, as a network of flows. Use a landscape's spatial qualities to inform criticism. Does the space embrace the visitor? Or does it keep them moving? Draw the space in plan and in section. Do the place's spatial qualities cohere? Do the landscape elements work together to contribute to a legible space? In a similar way, use health to inform critique. Is the health of the space decrepit or is the space healthy and lively? Examples of the former might include an urban office plaza only used by the wealthy or a creek bank devoid of riparian vegetation. Examples of the latter might include an active playground or greenbelt. The twin frameworks of space and health can organize dialogue with a place by incorporating critical elements within them.

Second, human nature being what it is, given limited time and attention spans, it may be necessary to focus on a particular aspect of a place and use that aspect to assess its health. Critical inquiry may need to choose – a place's racial history, the threat of tsunamis or the mutualism of mycorrhizae. At times, this decision is obvious; the client has hired the design team to address a landscape's lack of activity, for instance. At other times, it is difficult to tell what landscape element might be the most important. One potential focus that embraces both content and context and has not received enough attention in traditional site analysis is the placement/displacement of people related to social and spatial equity. Displacement is a spatial concept, the removal of something or someone from one area to another. It can be a positive change or aspirational if movement is initiated by the mover or it can be negative if movement is forced upon the mover. Forced displacement includes the exclusion of people experiencing homelessness from an urban plaza, the elimination of the habitat of an endangered species or the movement of families to another neighborhood through redevelopment. As the designer contemplates a landscape critically, an analysis of displacement(s) structures the selection of significant elements and flows. Displacement is an integral part of the landscape, shaping its configuration, form and symbolism. We must be open to existing inequities, asking how past forces affect the form and movement of people in spatial design.

To engage in landscape dialogue does not necessarily require one to evaluate displacement as a "conscious pariah," but to do so critically does. It takes training. We explicate displacement in the landscape further through a dialogue with violence.

DISPLACEMENT AND VIOLENCE

The experience of displacement changes the character of a place. In an extreme breakdown of relations, violence occurs. While violence can be dismissed, hidden, ignored and paved over, it cannot be completely removed from the landscape. It persists, seeping into the soil and the movement of people through a place. It is found in the demeanor of the disenfranchised and the strategic machinations of power. Event-based violence, that of battles, riots and heroes, begs for memorialization in the landscape, while the more pervasive daily violence of rejection and displacement may be unacknowledged.

Monuments to violence tap into a collective memory or an assumed collective memory of the dominant culture to explain the event and the landscape. Not all violence gets commemorated. Kenneth Foote states that monuments tend to either obliterate violence and its memory or memorialize it.[26] Foote describes society's need to forget violent acts (particularly those killing innocent people) as an attempt to categorize this violence, not as dissent, but as a deranged aberration to ordinary democratic dialogue.[27] But violence, whether perpetuated by the State as in Selma, Alabama or unleashed by a domestic terrorist as in the Oklahoma City bombing, changes a community and the landscape in ways that require acknowledgement. Conflict does not disappear amidst rational debate. It can fester. Monuments in the landscape – to war, to heroism, to a nation – can operate as foils to the critical, becoming the most uncritical of landscape elements. What is it we are remembering – the hero on his white horse or the people he vanquished? Monuments emphasize a dominant story, while obscuring the displacement of people and the landscape.

Often violence in the landscape is not an event but a way of life. In contrast to the current stability of some countries and their borders, the West Bank of Palestine/Israel (*ad-Ḍiffah al-Ġarbiyyah* in Arabic) is a territory of shifting claims, powers and spaces. It is the territory of ancient civilizations of Middle Eastern cultures with cities like Jericho some of the oldest in the world. Significant holy sites of Judaism, Christianity and Islam sit within its borders. Jordan claimed the West Bank after World War II before it was captured by Israel during the Six Day War of 1967.[28] Since then, Israel has colonized the territory through Jewish settlement, military infrastructure and a system of apartheid (meaning "separate" in Afrikaan). While most countries of the world label the West Bank as "Occupied Palestinian Territory," the Israelis control much of the movement and discourse within the West Bank.

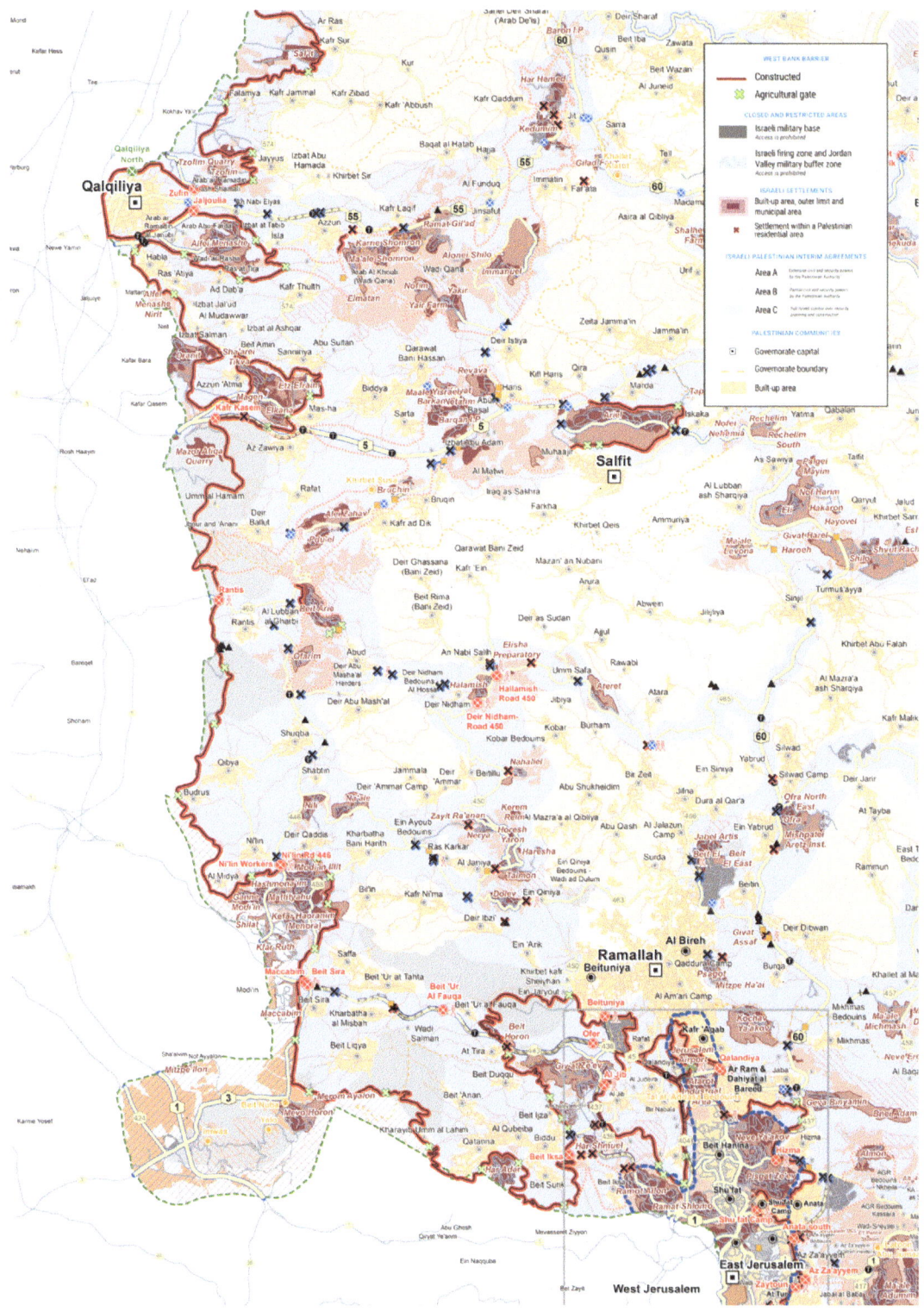

6.8 Portion of United Nations map of settlements in the West Bank, Palestine. Red line is the West Bank barrier.

Eyal Weizman, a British Israeli architect, in his book *Hollow Land* explicates the militaristic structure of the landscape of the West Bank – the "architecture of occupation" as he calls it.[29] The architecture is designed to separate Israelis from Palestinians. While settlements of the West Bank by Jewish settlers have been portrayed as peaceful, even working into the language of the Oslo Accords, Weizman says the purpose is to fragment or partition the landscape to establish apartheid (segregation). There are three dimensions to landscape apartheid, all related to national security (for Israel): (1) horizontal displacement of Palestinian towns and neighborhoods as separate from Jewish settlements through walls and control of flows; (2) Jewish settlements linked together through highspeed highways and tunnels, facilitating movement to Jerusalem and the rest of Israel; and (3) Palestinian towns connected by a transportation system with a series of controlled points.

> To control a space, you need to create differentiation in speed of movement. When you put Israeli colonies on highway, you accelerate their movement through space. In the same way, in every twist and turn of the terrain, Palestinians would encounter a border, a checkpoint, a fence, a valley that they cannot cross.[30]

Vertical displacement occurs as Jewish settlements shove the Palestinians from the high places of the West Bank.

In an attempt at reducing violence, in this case, that of Palestinian suicide bombers entering Jerusalem, Israel replaced it with a different kind of violence. Weizman describes Israeli architecture of walls and checkpoints as a "slow violence."[31] The walls, bifurcated highways, checkpoints and surveillance towers of the Israeli occupation were designed to gradually make the place unlivable to Palestinian inhabitants. Since the Six Day War of 1967, the Israeli military has torn down houses, fought skirmishes over villages and encouraged Jewish settlement of the high points, gradually, slowly shifting the West Bank from a Palestinian homeland into a stratified region of haves and have-nots.

The West Bank may be a dramatic example of displacement and slow violence but it would be difficult to find a landscape *not* subject to past violence. In the United States and Australia, for instance, the colonization of the landscape exterminated large numbers of indigenous peoples and pushed remaining peoples into less productive areas of the country (reservations). In South Africa, a system of apartheid/separation kept

6.9 This steel wall protects an "Israelis only" road that cuts through Palestinian land, separating people from their neighbours and their farms.

native Africans removed from white Afrikaaners who operated within different social, economic and political networks. In India, a rigid system of caste places people into categories based on worth to prevent the lower castes from contacting the higher castes. Each of these systems relies on direct and indirect slow violence to keep certain people from accessing needed social and physical resources.

FORENSIC ARCHITECTURE

Weizman's most critical idea is that *matter is political*. Existing physical landscape elements are the result of deliberate interventions in the structure of the place – concrete, reinforcing steel, the walls of glass staring down at the Palestinian homes. And if that is so, then the "reading" of matter, a critical dialogue with matter, becomes a skill to be learned and applied to confront histories of violence.

Police forensics can help find a killer and bring him/her to justice, assist in the quelling of rebellion or expose injustice. It is most often a legal tool of the State. Weizman asks the question: can forensics be used against the State to expose hypocrisy, inequity and slow violence? "Turning forensics against the state is essential because of the intertwined nature of state violence, which, as previously mentioned, is both violence against people and things and also against the evidence that violence has taken place at all."[32]

Weizman and his colleagues work in "frontier zones" like the West Bank, the Mediterranean Sea, Somalia and west Pakistan, where the State has less control, as he describes it, "the lawless battlefields of our colonial present."[33] The idea behind forensic architecture is that buildings speak, if we listen. Violent events persist in the stucco walls of a building or the shallow wadi behind the village. As the Palestinian poet Najwan Darwish elucidates in *Jerusalem*:[34]

> The murdered hum their poems on the hills
> and the rebels reproach the tellers of their stories
> while I leave the sea behind and come back
> to you, come back
> by this small river that flows in your despair.

Weizman compares the forensic architect to the geologist – both examine cracks and disturbances in the (sub)structure to reveal underlying tensions and forces acting upon it. A road cut through a hillside reveals layers of rock from prior millennia that tell a story. A crack in the wall tells of past construction techniques and events.

> Cracks are material events that emerge as the result of force contradictions. They progress along paths of least resistance, exploiting and tearing through different material substances where the cohesive forces of aggregate matter are at their weakest. Each crack is a unique result of a specific disposition of a force field and material irregularities on the micro level.[35]

It is not just architecture which speaks. Early in my career, I worked for a wetland ecology practice. A team of soil scientists, botanists and environmental scientists and I (the designer), tromped through swamps and marshes of the Pacific Northwest, delineating the boundaries of wetlands according to the mandates of the U.S. Clean Water Act. We drew up restoration plans. We rearranged the flow of degraded streams. We also testified in court. It is not legal to discharge any pollutant from a point source into navigable waters of the United States (Clean Water Act, Section 404). In one case in California, the Environmental Protection Agency sued a rancher who had allegedly filled vernal pools, a rare, and usually dry, wetland that hosts endangered vernal pool fairy shrimp. Working with the Environmental Protection Agency's lawyers and under the watchful eye of the opposition, we excavated a three-meter trench through the area of supposed fill to unearth a clear soil profile. The soil bore witness to recent wetlands: two meters of a sandy loam without structure, then a thin layer of clay over a clay loam with clear evidence of shrink and swell. But the localized evidence was not enough. We had to contextualize the soil profile and examine the rest of the ranch. As the landscape architect, my role was to complete a topographic documentation of past and present surfaces – a model of the rise and fall of the clay lens that indicated (former) pools. We wandered the landscape with a measuring rod and a laser level, sticking it down holes, along sloughing creek banks, creating a composite model of the past and present surfaces of the ranch. Wielding a 20-foot measuring rod in high winds to make accurate readings while being filmed by the opposition's lawyers was landscape critique, in uncovering that which is hidden, in documenting transgression and revealing its effects from the positionality of wetland advocate.

While not everyone has a knowledge of soils to assess a profile to the degree of offering legal testimony, designers can learn enough soil science, nutrient dynamics and wildlife ecology to make quick assessments of

6.10 Sandy soil profile, Virginia.

ecological processes by using United States Department of Agriculture soil maps, Munsell Color charts and formal decision trees. That which is below the surface or in the past does not have to remain invisible or unknown. Forensic historians uncover burial grounds, offering evidence to past atrocities and genocide, as well as offering alternative histories based on evidence in the landscape.

PRAXIS

Imagine a young landscape architect faced with a new project for a large retail center to be built on a wetland on the outskirts of town. The project would lead to a few (minimum wage) jobs in the community. It would also destroy functioning wetlands and add more car trips within and beyond the community … two impacts not recognized by the majority of the community. One option in the design work would be to move forward with designing the retail center landscape as proposed. To engage with the client, in this case the property owner, the architect and the engineer, and swallow one's own concerns. But the acceptance of a limited role in the project hinders learning about the space, the functioning processes, the social milieu of the development. This is the path of listening without speaking.

Now imagine this scenario with a young landscape architect who embraces landscape critique. Dialogue begins with questions (which may be interpreted as dissent). These questions arise from three sources:

1. Local people, tribes, communities.
   Whose place is this? What claims do other cultures have on this place? What is their vision of this place?
2. The client (in the broadest sense).
   Why is this development placed here? What is gained? What is lost? Whose voice is not being heard in meetings? Are there alternative locations for the project?
3. The landscape.
   The landscape has its own set of questions, often revolving around health. A partial list: What was this place like before urban development? Why was this place changed? What do the plants tell us about the health of the soil? What does the surface of the landscape say about porosity or imperviousness? What spaces exist as gathering areas, as sinks, as rooms? What spaces open up to the surroundings?

Can existing buildings be
re-used to house planned
activities and uses?

How can this space be
re-connected to the river?
Visually? Ecologically?

How does the levee sepa-
rate the landscape from
the river? Would this work
in a 200 year flood?

Why are riparian trees in
the pubilc right of way
being cut down?

Who owns the site? Who occupies
the site? Who used to occupy the
landscape in this area? Do they
still have a voice in the place?

6.11 Questions asked about a light industrial site to be redeveloped.

Landscape critique asks questions that relate one landscape element to another. So instead of asking a specific question of circulation patterns, ask how people's arrival relates to their experience of the landscape. Or examine the conflicts inherent in different modes of transport to suggest potential resolutions.

The answers to these questions may inspire the designer to refuse to work on the project. But if they stick around, asking direct questions leads to a deeper understanding of place, particularly as the questions receiving the most pushback are often those issues that are, ironically, most visible but most assumed (i.e. existing speed limits through town). And these issues are also the most likely to require an uncovering, a make-visibleness, a deeper understanding. It may be that the developers acknowledge the importance of wetlands (from an ecological and legal perspective) and have already planned for extensive mitigation, while the issue of increased traffic and length of trips has not been considered.

As in this imagined scenario, many of the landscapes we encounter in our professional and academic lives will not bear visible marks of a violent history, like in the West Bank. Yet, even most mundane places contain palimpsests of violence, both fast and slow. We acknowledge, collectively, that a lack of knowledge regarding those non-places in our lives may end up hiding this violence. How then is forensic architecture practiced?

Forensic architecture is an acknowledgement that buildings and landscapes are dynamic. Although slower than the growth of plants, building materials wear down, crack, blister, shrink and swell. Present patterns and forms can be used to make estimates of how the landscape has changed over time. This change can be an event, as the deposit of fill in a wetland by bulldozers, or it can be very gradual, as in erosion (see Chapter 1). Like Weizman, we are interested in forensic analysis not (strictly) as a documentation of landscape change, but as a process of uncovering displacement or even violence. This might be a dramatic event itself, the cutting off of a statue's foot, but is more likely to be the presence or absence of landscape elements that require difficult questions.

1. Identify

Identify the story of a specific landscape – what happened here? A wetland has been filled, a people displaced, a new building obscures the view of prior residents. Is someone no longer here? Construct a timeline of past events focusing on the transition points between peoples, zeroing in on moments of change or conquest. Identification will include the invisible elements of a place, including that which is underground, such as soil, utilities, human remains and groundwater. Identification will also include the existing materials and their properties – where do they come from? Where will they go after their life cycle ends?

At the confluence of the Columbia and Yakima Rivers in the Pacific Northwest of the United States, a local non-profit agency proposed building an interpretive center as a gateway to the Hanford Reach National Monument, a museum to celebrate the last free-flowing stretch of the Columbia River. A visit to the site discovered a hummocky micro-topography composed of undulating mounds west of Interstate 182. I considered a number of natural processes that could have generated this topography but in consult with several long-term residents, they graciously pointed me to the truth. The hummocks were the result of local residents digging up Indian artifacts – pots, arrowheads – to display them

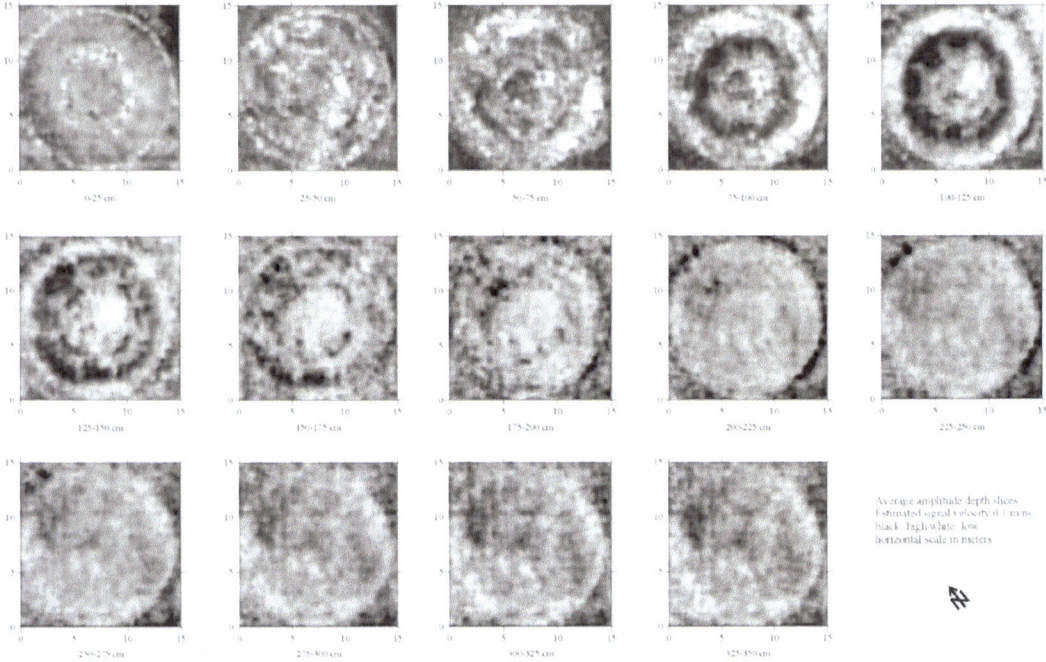

6.12 Ground penetrating radar output showing different densities of a crypt in St Paul, Minnesota taken at slightly different depths.

in their homes, most of the activity taking place in the 1950s and 1960s. We engaged with local tribes, the Nez Perce and Yakima, to assess the landscape and begin formal negotiations for this space. We conscripted an archaeological firm to use ground penetrating radar in this landscape to uncover the density of underground artifacts and to make sure there were no human remains. Eventually, the Yakima Nation nixed the location as too culturally sensitive to accommodate a museum or any other development.

2. Re-create

Re-create the scene of past events – violent or significant or communal. What did this place look like before the event(s)? What was the nature of the violence that took place? Weizman re-creates an event using standard architectural tools, such as three-dimensional rendering software and GIS. But the human body can also be used.

In 2018, police in Sacramento, California, responding to a call complaining of a car break in, chased a young black man through a residential neighborhood before shooting him 20 times when he pulled out his cell phone. His name was Stephon Clark and he was killed in his grandmother's backyard. Sacramento's African American community assembled to protest his death at the hands of the police, leading to a series of marches that culminated around his funeral. Beginning at Sacramento's City Hall, the protesters against police violence walked west into the former West End neighborhood, before entering an on-ramp of Interstate 5 and blocking traffic on the highway during rush hour. Their path was deliberate. The West End used to be occupied by a multi-ethnic community of Chinese Americans, Japanese Americans and African Americans in a neighborhood of small shops, second story residences and jazz clubs. In the 1950s, the City forcibly displaced ethnic minorities from this vibrant community in the name of redevelopment. Many of the lots with torn down buildings sat vacant for decades before eventually becoming office towers, the historic Old Sacramento, an interstate and a new basketball stadium – a more "acceptable" entrance to the city after crossing the Sacramento River. We, as in collective humanity, are constantly re-creating stories of the landscape in an effort to assert our own visions of how things have come to be. Sacramento's

6.13 A 1962 aerial photo of the West End showing new vacant lots at the entrance to the city and the future site of Interstate 5.

African American community *re-created* through dissent a dynamic, multi-ethnic community to reveal the historic removal of black people from the West End.

3. Tell a story

Present the situational violence as a part of an ongoing landscape story. Bring this presentation to the public forum, to the design team, to the client. The intent is not to physically restore these spaces to their former glory. It is that this landscape critique must be examined before spaces of landscape healing can be proposed and implemented.

In the case of the West End in Sacramento, a certain type of commerce, in the form of office complexes, hotels, restaurants and a basketball stadium have replaced residents, because of a certain urban aesthetic necessary for the maintenance of white privilege. Jesus Hernandez uses "real estate security maps" showing redlining of the city to illustrate systemic racism that shaped the current composition of Sacramento's spatial distribution.[36] Efforts to bring back residents to a place that is "dead" at night, after office workers leave, cannot proceed without recognizing the decades of twentieth-century violence that made the current configuration possible.

Sometimes the presentation of slow violence can be straightforward, as in the redlining maps that form a clear visual statement of racism perpetrated by banks and local economies. But other times, presentation is less clear; the slow violence is difficult to understand. Original, past violence against Native Americans in California, for instance, are recorded in documents, diaries and land records of the nineteenth century, but these accounts are incomplete and often from the perspective of the conquerors. The California tribes of Nisenan, Patwin and Miwok may be hesitant to share their own stories as past attempts have been appropriated, distorted or relegated to a curiosity in the recent past. Thus, presentation, putting something on paper as a critique, must be done sensitively with open lines of communication and the maintenance of ambiguity. In these cases, art offers a way of showing situational relationships, imbalances of power, while preserving mystery in the landscape.

NOTES

1. Story compiled from Juliane Schwarz-Bierschenk, "Monumental Discourses: Sculpting Juan de Onate from the Collected Memories of the American Southwest" (PhD thesis, 2015), https://epub.uni-regensburg.de/31523/; Michael

Trujillo, "Oñate's Foot: Remembering and Dismembering in Northern New Mexico," *Aztlan: A Journal of Chicano Studies* 33, no. 2 (September 1, 2008): 91–119.

2. Simon J. Ortiz, "Time as Memory as Story," Poetry Foundation (Poetry Foundation, May 6, 2021), https://www.poetryfoundation.org/, https://www.poetryfoundation.org/poems/53442/time-as-memory-as-story.

3. *Santa Fe Reporter*, "Proud Actions," September 9, 1998, 5.

4. Dale Becker, Whisper Camel, Cory Parker, Charlie Scott and Ints Luters, "Wildlife Crossings within the Cultural Landscape of the Flathead Indian Reservation: U.S. 93 from Evaro to Polson, Montana" (International Conference on Ecology and Transportation, Seattle, Washington, 2011).

5. Schwarz-Bierschenk, "Monumental Discourses," 124.

6. Alison Fields, "New Mexico's Cuarto Centenario: History in Visual Dialogue," *The Public Historian* 33, no. 1 (2011): 44–72, https://doi.org/10.1525/tph.2011.33.1.44.

7. Trujillo, "Oñate's Foot."

8. Stan Alcorn, "Oñate's Foot," *99% Invisible*, December 4, 2018, https://99percentinvisible.org/episode/onates-foot/.

9. Carole Wade, "Using Writing to Develop and Assess Critical Thinking," *Teaching of Psychology*, February 1, 1995, https://doi.org/10.1207/s15328023top2201_8.

10. Julie Fisher Robertson and Donna Rane-Szostak, "Using Dialogues to Develop Critical Thinking Skills: A Practical Approach," *Journal of Adolescent & Adult Literacy* 39, no. 7 (1996): 552–556.

11. Margaret McAvin, Elizabeth K. Meyer, James Corner, Hamid Shirvani, Kenneth Helphand, Robert B. Riley and Robert Scarfo, "Landscape Architecture and Critical Inquiry," *Landscape Journal* 10, no. 2 (September 21, 1991): 155–172, at 167, https://doi.org/10.3368/lj.10.2.155.

12. J. Barr and Morwenna Griffiths, "'Training the Imagination to 'Go Visiting'," 6, https://www.academia.edu/11358746/_Training_the_imagination_to_go_visiting_.

13. McAvin et al., "Landscape Architecture and Critical Inquiry," 163.

14. bell hooks, *Teaching Critical Thinking: Practical Wisdom* (New York, NY: Routledge, 2009), 185.

15. Hannah Arendt, *Lectures on Kant's Political Philosophy* (Chicago, IL: University of Chicago Press, 1989).

16. As quoted in Lisa Jane Disch, *Hannah Arendt and the Limits of Philosophy*, 1st edition (Ithaca, NY: NCROL, 1994), 157.

17. June Manning Thomas, "The Minority-Race Planner in the Quest for a Just City," *Planning Theory* 7, no. 3 (November 1, 2008): 227–247, https://doi.org/10.1177/1473095208094822.

18. Hannah Arendt, *The Jew as Pariah: Jewish Identity and Politics in the Modern Age* (New York, NY: Grove Press, 1978).

19. Arendt, as quoted in Disch, *Hannah Arendt and the Limits of Philosophy*, 176.

20. Disch, *Hannah Arendt and the Limits of Philosophy*.

21. M.M. Bakhtin, *The Dialogic Imagination: Four Essays* (Austin, TX: University of Texas Press, 2010), 287.

22. Elizabeth K. Meyer, in Margaret McAvin, Elizabeth K. Meyer, James Corner, Hamid Shirvani, Kenneth Helphand, Robert B. Riley and Robert Scarfo, "Landscape Architecture and Critical Inquiry," *Landscape Journal* 10, no. 2 (1991): 155–172.

23. See classic works such as David Harvey, *Social Justice and the City*, Revised edition (Athens, GA: University of Georgia Press, 2009); Don Mitchell, *The Lie of the Land: Migrant Workers and the California Landscape* (Minneapolis, MN: University of Minnesota Press, 1996); Denis Cosgrove, *Social Formation and Symbolic Landscape*, New edition (Madison, WI: University of Wisconsin Press, 1998).

24. McAvin et al., "Landscape Architecture and Critical Inquiry."

25. Elizabeth K. Meyer, in McAvin et al., "Landscape Architecture and Critical Inquiry."

26. Kenneth Foote, "On the Edge of Memory: Uneasy Legacies of Dissent, Terror, and Violence in the American Landscape," *Social Science Quarterly* 97, no. 1 (2016): 115–122, https://doi.org/10.1111/ssqu.12259.

27. Kenneth E. Foote, *Shadowed Ground: America's Landscapes of Violence and Tragedy*, Revised edition (Austin, TX: University of Texas Press, 2003).

28. To summarize in one paragraph the expansive, chaotic and controversial history of the West Bank would be futile. For a book-length history of the Israeli–Palestine conflict as it relates to the landscape, read Meron Benvenisti's *Sacred Landscape: The Buried History of the Holy Land since 1948* (Berkeley, CA: University of California Press, 2000). And for an examination of life lived in present day West Bank, try Ben Ehrenreich's *The Way to the Spring: Life and Death in Palestine* (New York, NY: Penguin Books, 2016).

29. Eyal Weizman, *Hollow Land: Israel's Architecture of Occupation* (London: Verso Books, 2012).

30. Special series, dir. 2014. *Eyal Weizman: The Architecture of Occupation*. Rebel Architecture. Al Jazeera, https://www.aljazeera.com/program/rebel-architecture/2014/8/12/eyal-weizman-the-architecture-of-occupation.

31. Weizman, *Hollow Land*.

32. Eyal Weizman, *Forensic Architecture: Violence at the Threshold of Detectability* (Princeton, NJ: Princeton University Press, 2017), 64.

33. Eyal Weizman, "Introduction: Forensis," in *Forensic Architecture*, ed. Forensis (London: Sternberg Press, 2014), 11, https://www.sternberg-press.com/product/the-architecture-of-public-truth/.

34. Najwan Darwish, "Jerusalem," trans. Kareem James Abu-Zeid, 2012, https:// www.poetryinternational.com/poets-poems/poems/poem/103-22183_ JERUSALEM.

35. Weizman, *Forensic Architecture*, 53.

36. Jesus Hernandez, "Redlining Revisited: Mortgage Lending Patterns in Sacramento 1930–2004," *International Journal of Urban and Regional Research* 33, no. 2 (June 1, 2009): 291–313, https://doi.org/10.1111/j.1468-2427. 2009.00873.x.

CHAPTER 7

Landscape dissent: transgression, protest and the body

In the morning hours of March 7, 1965, a double row of serious-minded people walked toward the Edmund Pettus Bridge in Selma, Alabama. It was the beginning of a 54-mile march to Montgomery, the state capital, to bring attention to the systematic denial of voting rights for black people. The Edmund Pettus Bridge rises from downtown Selma to the south up and over the Alabama River before descending to a small group of commercial buildings and open space on the other side. The purpose of the bridge is to move cars between rural parts of Western Alabama and Selma. It is a landscape of transportation.

The African Americans, led by the young Student Non-violent Coordinating Committee (SNCC) activist John Lewis and Hosea Williams of the Southern Christian Leadership Council (SCLC), marched on to the bridge, ascended its eastern side and crested the top to see a phalanx of state troopers waiting on the other side. They descended the apex of the bridge and stopped. Major John Cloud of the Alabama state troopers ordered them to disperse, giving them two minutes. Grainy photographs of the standoff show the vulnerability of the protesters' bodies placed in the open facing the patrollers. Williams asked for a word with the Major but there was nothing he wanted to talk about. It was a tense moment, broken by the order to advance. What followed was broadcast on national television, interrupting the day's programming: policemen knocking protesters to the ground then beating them with billy clubs. Horses stepping over prone bodies on the pavement. The first 20 African Americans in line were quickly bleeding on the pavement and the rest of the line turned to flee. Andrew Young, an SCLC organizer, recalled: "The police were riding along on horseback beating people. The tear gas was so thick you couldn't get to where the people were who needed help."[1]

DOI: 10.4324/9781003158943-8

7.1 Marchers in Selma, Alabama on March 7, 1965, crossing the bridge.

In past chapters, we discussed landscape as a network of relations, relations bound up with power. We talked of visible relations, some obvious, some difficult to encounter because they were assumed, as well as invisible relations, hidden in the past or underground. However, with hidden relations we immediately run into a problem: How does one "see" invisible relations that might inform landscape character if they are invisible? This is part of dialogue, particularly dialogue between those with power and those without. In dialogue, we use dissent to bring something up that does not want to be brought up. We use it to not just criticize, but even more fundamentally, to reveal, to add something to an agenda that was not there before.

A number of years ago, I visited the Edmund Pettus Bridge on a tour of the South led by a Civil Rights group called Museum without Walls. We drove to Selma in the early morning hours and met with two sisters who had survived the protest as teenagers. They gave stark accounts of that day … "the beating and beating and beating." When we came out of that meeting, a little disoriented, I wanted to see the bridge – to walk up one side and down the other, like John Lewis and the protesters had done. Suzanne Lacey, the director of Museum without Walls and our guide that day, had another idea. She led us north along Water Avenue to an older building without first floor windows, a blank wall. A diminutive woman wearing an African head dress came out of the door and commanded us to line up against the wall. We entered the building, crowded into a cramped entry. She detained one of us, separating the group. For the next 45 minutes, she led us through the journey of a slave from capture to a slave ship to a slave market, whispering to me "We are taking your children, your family," along the way. It was a harrowing experience. The tour guide had to end the simulation early, so many of us had either abruptly bowed out or looked catatonic. We wandered the gallery of the Southern Slave Museum after our experience, pondering our own mixture of horror and privilege.

Until that point, the ramifications of slavery had been invisible to me. It was so long ago (although, even that statement is not quite true, according to the Global Slavery Index).[2] If I had gone straight to the bridge, I would have missed the deeper connection that voting rights had to the silencing of voices over centuries. I would have misunderstood the role of slavery in the initial development of the town as a trading center. I would have never searched for the earthen embankments that once protected the town's iron forge during the Civil War. Each landscape harbors a past of conflict that informs the present.

7.2 Edmund Pettus Bridge over the Alabama River.

Conflict can be direct conflict of argumentation, resistance and violence or it can be indirect conflict of exclusion and disinvestment. What then reveals conflict in the landscape – the hidden "power geometries" at work? It is struggle, as in the small case of a re-enactment of a slave experience. It is dissent, as in the crossing of a bridge by an oppressed people. This chapter expands on the last chapter on landscape critique to tackle the agency of the "powerless" to make visible structures embedded in the landscape.

DISSENT, LANDSCAPE AND MEMORY

To find out how dissent reveals hidden structures in the landscape requires a further examination of what happened on the Edmund Pettus Bridge. First, why did the state patrol react with such violence against the marchers crossing the bridge? It relates to geographic scale. In the history of the Civil Rights movement in America's South, a pattern is clear. Black residents would agitate for change. The local sheriff and white residents of Montgomery, Alabama, Albany, Georgia, or Little Rock, Arkansas would push back with indirect and direct measures, some successful and some not.[3] At a certain point, the dissent and pushback reached a level which demanded notice by the national media. Martin Luther King Jr., a national figure in the movement, would arrive, give a speech, exhorting the protesters to non-violent protest. For local law enforcement, this changed the dynamic ... now, someone from out of town was riling up "our negroes."[4] A new white transplant to Albany, Georgia found "both white and colored to be friendly and sincere, living together in harmony until 'outsiders' ... troublemakers, and promoters of hate and strife, came to our city."[5] A local issue (which could be controlled) morphed into a larger

Civil Rights issue (which could not). And if it got to that level (the thinking goes), real change might happen: integration of schools, black and white people eating at the same restaurant, a re-evaluation of humanity and their own place in it, a change in everyday life. Local sheriffs were quick to blame national personages but slow to see how their local actions affected national sentiment, seemingly unaware of the broadcast of their actions nightly and the link between those broadcasts and Civil Rights legislation.

Hundreds of cars crossed the Edmund Pettus Bridge daily; this movement of pedestrians was different. The escalation of the protest in Selma, by crossing the bridge onto a state highway on the way to the capitol required an escalation in enforcement, from the local sheriff to the state police. That bridge over the river named for a white man, a former Confederate brigadier general, was a literal and symbolic escalation from the local to the state. Organized protest *revealed* the boundary between local and state, how the landscape (in the form of the bridge) constrained local African Americans to the local environs and away from state politics and decision-making.

Dissent exposes barriers to movement from local to global (justice). Each landscape of dissent is situated in its own place but connected to larger systems of oppression and other places of protest. To understand the landscape, the designer must dialogue with a neighborhood/landscape/town while in conversation with global issues of climate, race and economy. The dialogue is not just with the immediate surroundings but the larger world.

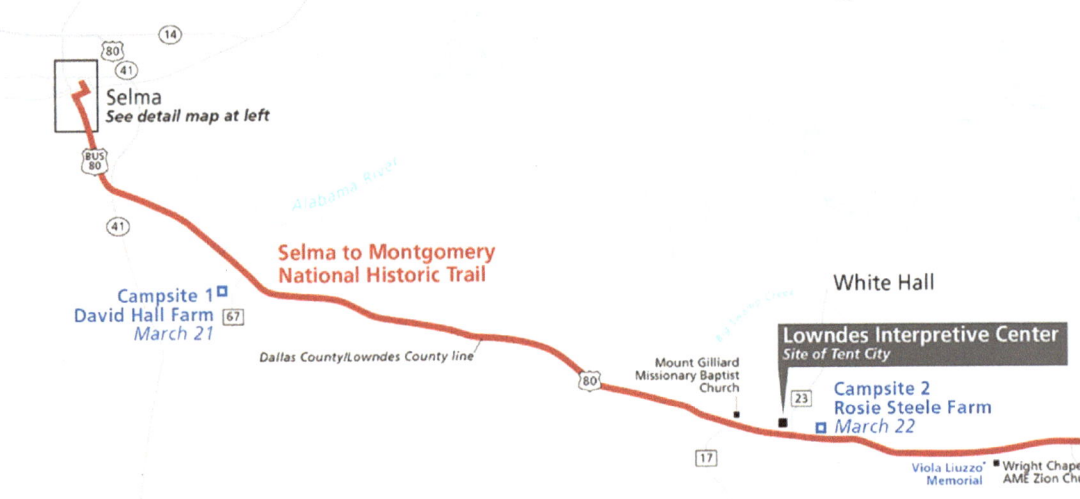

7.3 Map of Selma, Alabama and State Highway 80, the Selma to Montgomery National Historic Trail.

Second, the march across the Edmund Pettus Bridge challenged dearly held conceptions of time. White segregationists recalled a "fonder" time when black and whites "knew their place," when they lived in harmony with each other, showing proper deference.[6] That this fonder time never existed was not especially relevant to those harkening to a past utopia. Memory (of those in power) selects those events affirming a specific viewpoint and then reimagines this memory in the landscape.[7] In the midst of this, a landscape of dissent does not remain silent. It gave up the body of Emmett Till, a 14-year-old shot by white Southerners and dumped in the Tallahatchie River. More recently, it harbored the simmering discontent of black people against segregation and police militarization in Ferguson, Missouri and the streets of Minneapolis after George Floyd. Memories are long, even in the absence of direct experience.

In the early 2010s, Oregon State University asked Jones & Jones to examine four cultural centers serving ethnic student groups housed in temporary structures on campus. They had noticed a greater drop-out rate among ethnic minorities and desired four new places of on-campus support. We reached out to these students to involve them in the location and design of their cultural centers – places of refuge in the midst of a large Beaux Arts-inspired university campus that reflected a dominant (white) culture. At one point, our design team proposed relocating the Black Cultural Center closer to the center of campus on a small parcel of open grass, not big enough to

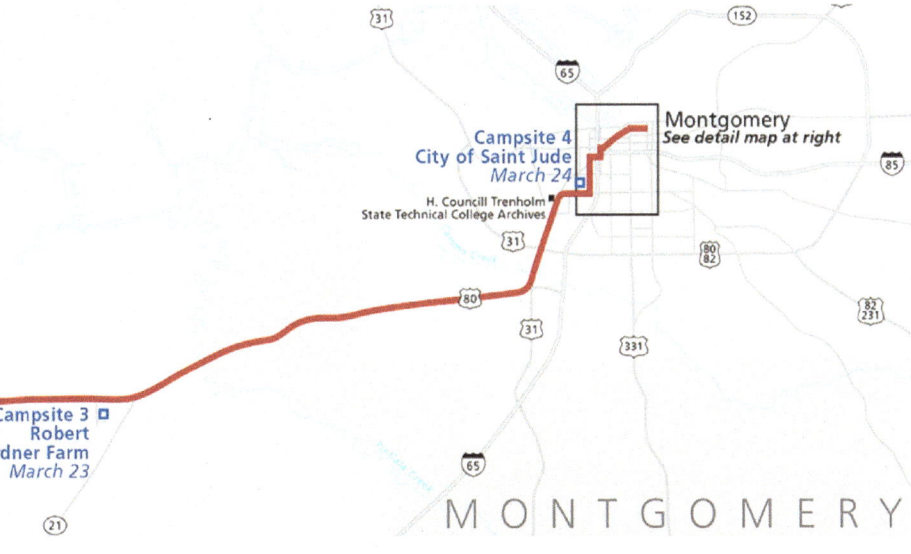

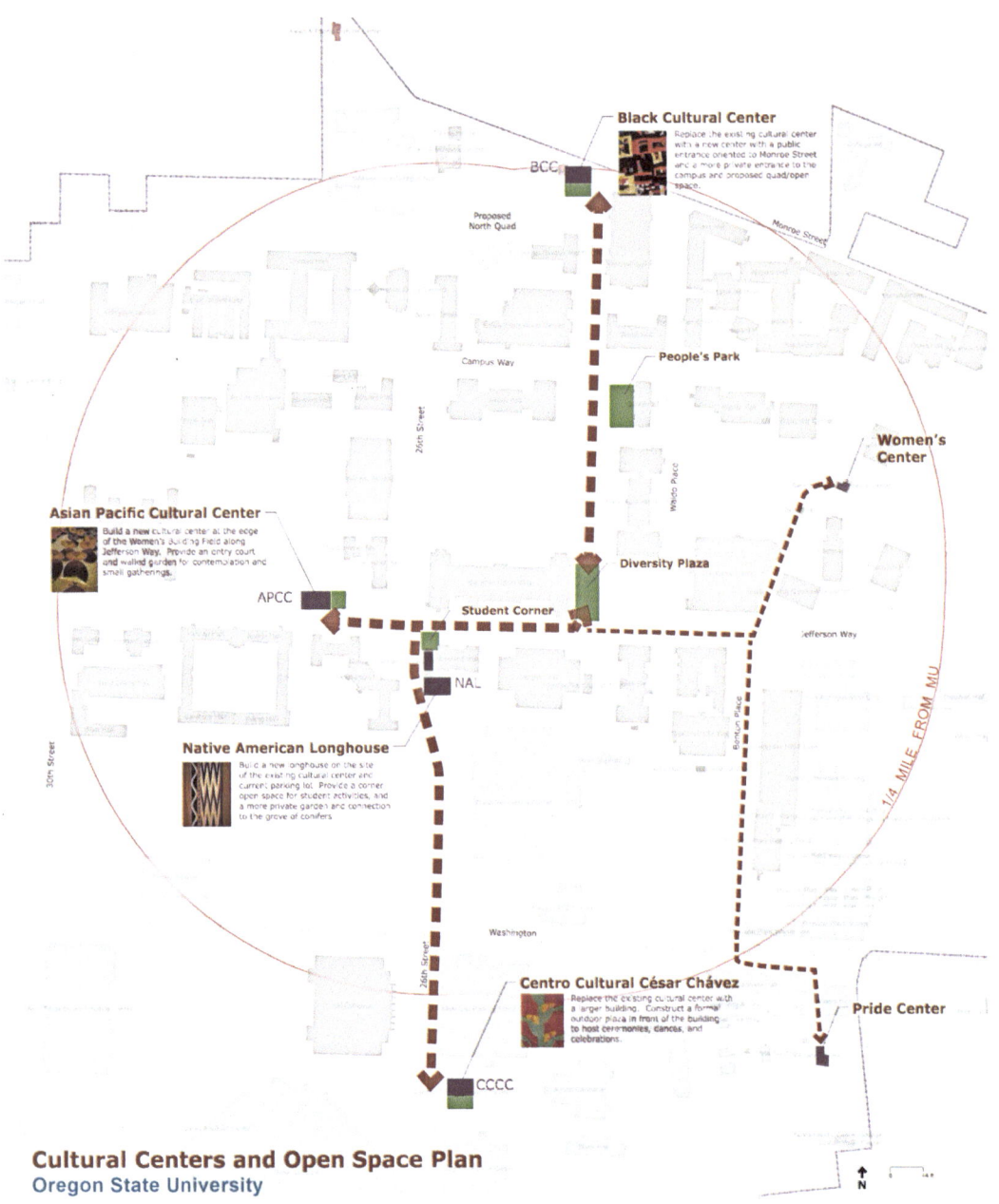

Black Cultural Center
Replace the existing cultural center with a new center with a public entrance oriented to Monroe Street and a more private entrance to the campus and proposed quad/open space.

BCC

Proposed North Quad

Monroe Street

People's Park

Campus Way

Women's Center

Asian Pacific Cultural Center
Build a new cultural center at the edge of the Women's Building Field along Jefferson Way. Provide an entry court and walled garden for contemplation and small gatherings.

APCC

Diversity Plaza

Student Corner

Jefferson Way

NA.

Native American Longhouse
Build a new longhouse on the site of the existing cultural center and current parking lot. Provide a corner open space for student activities, and a more private garden and connection to the grove of conifers.

1/4 MILE FROM MU

Washington

Centro Cultural César Chávez
Replace the existing cultural center with a larger building. Construct a formal outdoor plaza in front of the building to host ceremonies, dances, and celebrations.

Pride Center

CCCC

Cultural Centers and Open Space Plan
Oregon State University

N

7.4 Oregon State University preliminary location of proposed cultural centers with People's Park the site of historic black student protest.

be a campus gathering space but too big to be decorative. The black students rejected that alternative. Unknown to us (and apparently administrators and campus planners), the space was a crucial site to protest a lack of black professors, the 1969 Black Student Union walkout and a 1996 boycott and march.[8] These protests took place long before any of the students we met with arrived on campus. Yet the site was an active presence in their continued struggles, a reminder of their agency and power as they daily walked through campus. Removing that space, even to place their own cultural center, would have meant dis-remembering dissent.

Third, what happened on the Edmund Pettus Bridge happened within a legal context of oppression and control. The 1965 march across the Edmund Pettus Bridge was not legally sanctioned. It took place outside of or against existing laws.[9] Marchers had originally sought a permit from a local judge to walk from Selma to Montgomery, but the permit was not forthcoming. King and Lewis advocated waiting for the permit but local activists won out, deciding to march on March 7. That morning, Lewis reluctantly agreed to lead the marchers, along with Hosea Williams and other leaders of SNCC and SCLC.[10] The marchers walked without a permit, against systems of State control, and so, had to be put back into the delineated space of haves and have-nots, black and white.[11] They transgressed to call attention to voting rights.

Transgression is not necessarily dissent. Most transgression happens without a speaking part. But there is an aspect of dissent that is enlarged using the tools of transgression. Some dissent occurs under such rigid and sanctioned constraints, it is unlikely to be successful or noteworthy. For instance, protesters seeking a permit to protest on the Washington, D.C. mall, with its labyrinthian public spaces and overlapping jurisdictions, must agree to restrictions that dilute the influence of the protest.[12]

DISSENT IN THE LANDSCAPE

If the structures of the landscape are obvious, assumed, taken-for-granted, and alternatives are invisible, then dialogue becomes difficult. Who do we talk to? Who is not being heard? As designers, we find a way to make things visible. This may mean shifting the idea of "client" as we did in U.S. 93 (see Chapter 5). It may mean thinking of the landscape in a more comprehensive way. It may mean paying attention to dissent.

In legal terms, the dissenting opinion is the voice of the minority. In dialogue, dissent counters the dominant discourse, not just presenting an alternative viewpoint but spoken *against* that discourse. Dissent can be ephemeral, a protest song amidst marching people or a broken window of a shoe store. It can be constructive, calling attention to alternatives, or destructive, as in the case of the Twin Towers on 9/11, where the alternative voice gets drowned out in the horror of the event, a violent and catastrophic dialogue between particular ways of viewing the world.

In landscape dialogue, dissent speaks from a particular position, both vertical and horizontal. The dissenter speaks up to someone above themselves in a hierarchy. The dissenter speaks from the margins to a place in the center. This spatial component of dissension situates the designer as a critical part of the process of attending to histories. The designer assumes two roles in terms of dissent: (1) uncovering histories and markings of dissent in the landscape; and (2) using dissent to understand the invisible relations of the landscape.

Uncovering marks of dissent

The most "readable" dissent in the landscape is violence, either the violence of the State to quell dissent or the violence of the dissenter with a cause. There is dissent that inspires violence against it (as in Selma, Alabama). In China, student protests in April of 1989 centered on Tiananmen Square in Beijing. On June 4 of that year, the State responded with violence, shooting hundreds or thousands of demonstrators who were attempting to block the military's entrance into the square. Since that time, the Chinese government is determined to erase the event from the past and the present. Visitors to Tiananmen Square cannot see physical signs of the protests. The monumentality of the space is designed to celebrate, at a large scale, the national unity of China but it cannot permanently remove the memories of violence.

Violence, whether perpetuated by the State or unleashed by a domestic terrorist as in the Oklahoma City bombing, changes a community and the landscape in ways that require engagement. Conflict does not

7.5 9/11 Memorial from the edge looking down, New York City.

disappear amidst rational debate.[13] It can fester. Kenneth Foote describes society's efforts to construct monuments for innocent victims of violent acts as the categorization of this violence as a deranged aberration to ordinary democratic dialogue.[14] If violence of this kind stems from the uneven nature of power geometries and silenced people, then we are partially culpable (without removing any of the direct responsibility of the perpetrators). In the case of the Twin Towers in Manhattan, the community – in this case, New York City and the United States – honored the victims with two thoughtful and compelling voids, designed by Michael Arad and refined by Peter Walker Partners. The experience of moving across, through a field of oak trees to the edge of the water pouring over the edge, prepares the visitor for a reckoning with the void, with absence. Landscape dialogue of the "site" began with a reading of survivor and emergency personnel accounts and a willingness to embrace what, in many ways, is the antithesis of the two skyscrapers in the presence of the scooped out earth and debris.

In landscape dialogue, there will be moral arguments for distinguishing whether dissent should be celebrated as in the case of Selma, Alabama or condemned as in the Twin Towers. This can change over time. Dissent can begin as protest of a State or systemic oppression, only to eventually transform into an origin story for a community or a nation, i.e. the Boston Tea Party before the U.S. Revolutionary War or Mandela's prison cell in South Africa before the end of apartheid. Landscape dialogue establishes

7.6 Field of chairs at the Oklahoma City National Memorial by Butzer Design Partnership.

a conversation with dissenters, survivors, family members and victims to determine in what way honoring or condemning is appropriate to the space. One of the most compelling landscape installations, the Field of Chairs at the Oklahoma City bombing site, designed by Butzer Design Partnership, engages directly with the missing victims through a lighted, grid of chairs that include 19 smaller chairs for the children killed in the bombing.

Often, though, the State and designers ignore prior dissent in order to forget conflict. In these cases, landscapes of dissent are "designed" by the "people," locals, the disenfranchised, who wield their bodies to re-create a place (however temporary) rich with meaning. The challenge becomes learning the stories of these original designers, a practice closer to reading newspapers, digging through archives and talking to elders than a direct experience of the site itself.[15]

Dissent to make things visible

Dissent reveals taken-for-granted inequities, such as the protesters of police violence reinscribing the West End neighborhood in Sacramento (see Chapter 6). If, according to continental philosophers, the most powerful rules, limits and boundaries of society exist as assumed norms – the disenfranchised affirming their own rule through the practice of daily life – then those norms need to be revealed to question unjust practices.[16] Bourdieu calls this set of agreed upon norms of daily life "doxa." Doxa's power to organize social and material spaces derives from its assumed nature. Landscape critique may be the first step in making doxa visible, at least with respect to the spatial boundaries and constraints on behavior in public. For critique to move beyond the realm of discourse and into the realm of space, we need the diagnostic method of dissent.

Dissent serves as a method of dialogue with the landscape and local/global communities. Dissent is a form of speaking – protest – that has the potential to reconfigure dominant spaces in the landscape. As part of landscape dialogue, the designer participates in dissent to distinguish hidden boundaries and constraints. Dissent relies on transgression to do this. It was Georges Bataille, the French intellectual, who considered transgression, not as the opposite of law, but as the law fully realized.[17] Transgression reveals the constraint, boundary or order which it transgresses.[18] According to Chris Jenks:

> To transgress is to go beyond the bounds or limits set by a commandment or law or convention, it is to violate or infringe. But to transgress is also more than this, it is to announce and even laudate the commandment, the law or the convention. Transgression

is a deeply reflexive act of denial and affirmation. Analytically, then, transgression serves as an extremely sensitive vector in assessing the scope, direction and compass of any social theory.[19]

Each theorist teases out the reflexive nature of transgression, a transgression that reveals the qualities of the thing transgressed.

Each transgression crosses a boundary from periphery to center or from outside to inside, from exclusion to belonging. The geographer Timothy Cresswell describes transgression as a practice used to show what is normative.[20] Cities create limits and boundaries to include some people while excluding others, whether that's zoning, private property or diffuse spatial norms (doxa). But in doing so, cities create space for transgression. As Jenks puts it: "every rule, limit, boundary or edge carries with it its own fracture, penetration or impulse to disobey."[21]

Cresswell distinguishes between transgression and resistance with the idea of intentionality.[22] Transgression is unintentional; resistance is intentional. Transgression may reveal the structures of place, but its ephemeral qualities may never lead to transformation, particularly as its practitioners do not intend to transcend limits. The ephemeral quality of transgression suggests the temporal nature of de Certeau's "tactics" and the temporary movement through an ordered space not belonging to the walker.[23] This is transgression as "getting by" or "making do."[24] In contrast, resistance practiced intentionally amplifies the power of dissent, creating counter-hegemonic spaces, however temporary, revealing power calcified in the landscape.

transgression   dissent

7.7 Crossing a boundary-as-transgression to affirm a border and boundary-as-dissent in order to question its existence or placement.

Types of dissent as diagnostic

Dissent in the landscape varies, depending on the place and circumstances. *Trespass*, unwelcomed traversing of private space, is one type of spatial dissent. A trespasser may transgress private property without intent to dissent or resist, just to get where one is going. This can happen in "public" space as well. I have been accused of trespassing while taking notes on behavior at a public light rail stop, because I did not have a light rail ticket. An intent to ride the light rail is required to be in this public space. A trespasser or group of trespassers may transgress private space with an intent to protest existing spatial inequities, as in Safe Ground, a homeless group that camped on public and private property in Sacramento, California, to bring attention to the lack of housing options in the city.[25] This was dissent. By camping in those spaces, their trespass revealed the discriminatory rules applied to public property which stated a person could hike or bicycle in a park but could not sleep. They questioned the nature of public space and what is meant by "public." Trespass can be transgression; it can be dissent.

Similar to trespass, *movement* can be dissent. Effective and well-known protest movements of the past, such as the women's suffrage movement in Great Britain, the Civil Rights movement in the United States and the movement for Indian independence used marches to dramatize their pursuit of freedom. The ability to move, particularly across borders, corresponds to freedom.[26] Particularly within people groups whose movement has been curtailed, such as Palestinians or Uighurs, deliberate movement of large groups of people exposes the controls that bind them. On a smaller scale, some movements, as practiced by dissenters, run counter to established movement norms: going slow when you should be going fast and fast when you should be going slow, moving in a different direction or moving without the proper equipment.

If dissent can take the form of movement, it can also take the form of *cessation* of movement. Stopping, standing, sitting, sleeping, staying, occupying a place, particularly those places dedicated to movement or pass-through. The high priest of modernism, Le Corbusier, designed cities according to an idea of speed/success that relied on "rational, orderly, and law-abiding" citizens.[27] Any deviation in the rational or orderly prevents city movers from reaching their maximum speed, and thus, presumably, personal and corporate success. For the modernist city,

> no one trespasses in this city. The constant accidental social mixing which Le Corbusier sees as so wicked and chaotic in the old Paris …

is planned away, and with it the dangerous aspects of speed - dangers of social and cultural collision.[28]

Thus, stopping becomes something to plan away lest there be "social mixing." Stopping in areas devoted to speed becomes a dialogic tool of dissent.

The final method of dissent as a diagnostic tool is *assembly*, gathering in a space. We turn to Istanbul, Turkey to examine a landscape of crowds, dissent and the reading of that landscape. A dramatic sequence of protests that would eventually dominate international news for months began with a few "tree huggers" – local environmentalists upset over the removal of a retaining wall and some trees from the northern edge of Gezi Park, Istanbul. On May 27, 2013, they stood in front of the bulldozers, preventing further construction of the planned Armory/shopping mall. By night, they had set up tents and prepared for a long stand-off. The next day, city maintenance staff set the tents on fire. The violent response to dissent sparked further protests, more people gathered and the month-long event became an international incident. To understand the protests over Gezi Park, we examine the history and layers of meaning in the square.

Taksim Square in the Beyoğlu neighborhood of Istanbul, Turkey began as a hub of infrastructure, a node for the distribution of waters

7.8 The idea of the city as something to be tunneled through – Le Corbusier's perspective from *Radiant City*, 1933.

7.9 Protests in Gezi Park, 2013.

from Europe to the rest of the city. The Ottoman Turks of the nineteenth century constructed the Halil Pasha Artillery Barracks in the northern half of the square, followed by additional barracks, a military hospital and other public buildings in the district. These buildings, meant to modernize the Ottoman city, sat uncomfortably beside an Armenian cemetery, particularly as Turkey began to persecute the Armenians near the turn of the century, culminating in 1915 in Armenian genocide: expulsion from the army, forced marches to the desert and disappearances of Armenian intellectuals. After the tumult of World War I, a monument to the Armenian intellectuals killed was erected in 1919 in the square, only to be removed three years later.

When the new government of Mustafa Kemal Ataturk assumed power, the area was transformed to reflect a more cosmopolitan, secular Turkey.[29] They erected the Republic Monument, transforming the large country field into an urban square, the first modern space in Turkey. They hired Henri Prost, a French urban planner, to remake the Taksim district, culminating in the Beyoğlu Plan. The Armenian cemetery was razed. The artillery barracks were demolished, to be replaced by Gezi Park, a large, Western-style,

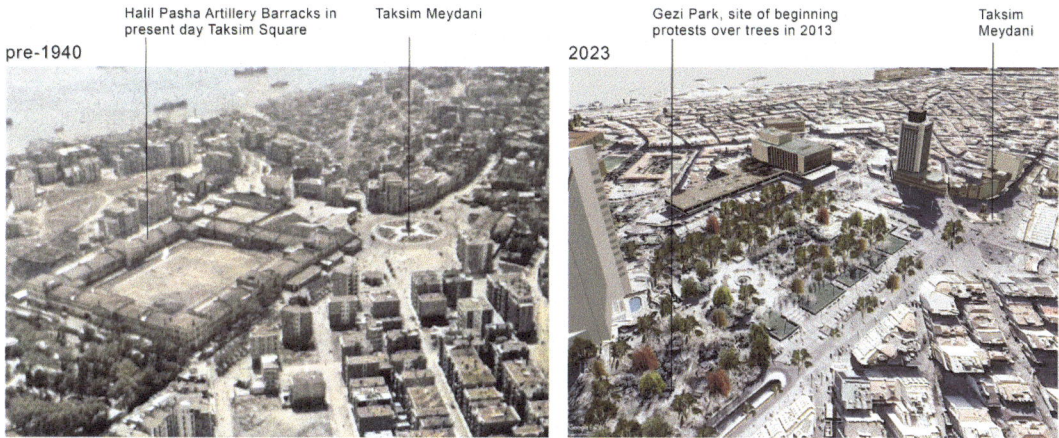

7.10 The gradual transformation of Taksim Square from artillery barrack pre-1940 (on left) to urban square (on right).

strolling surface of pavement and trees. The scale of Taksim Square, the larger space containing Gezi Park, approximately four city blocks, lends itself to the largest gatherings, whether political or celebratory. There is not another large open space in modern-day Istanbul like this.

The large space reflected a modern openness to secular thought, inspired by urban plazas throughout Western Europe. It hosted protests in 1977 for leftist labor causes, sponsored by the Turkey's Confederation of Revolutionary Workers' Unions. Around 500,000 people showed up to protest working conditions. Unknown assailants shot into the crowd killing 34 people, the Labor Day Massacre.

Layers of history built, covered, erased and remembered spread from the larger Taksim Square to Gezi Park. A palimpsest of infrastructure, memorials, military and protest. A space with multiple meanings and symbols. As the twenty-first century began, the government of Recep Erdogan, aware of the symbolic nature and centrality of the square, proposed to pedestrianize the space. The idea was to eliminate cars; make it walkable from end to end. Add a mosque. Transform it from a modern, secular space to a more singular space of (mis-remembered) Ottoman glory. However, attempts to erase the past (or pasts) are often met with resistance.

Whitehead and Bozoglu explain this process as the contestation of State-sponsored reconfiguration of space in the singular image of the State, rather than embracing its multiple dimensions.[30] Their term for past layers of history the State attempts to erase is "ground memories."

Ground memories are evident physical traces, presentations, and constructions of place histories that enable, influence, or inform human behaviors, beliefs, and senses of belonging. The sense of "ground" here is both physical and metaphorical. It is the earth and what is buried in it, what is or was known to be built on it, was erased from it or took place on it.[31]

Dissent and protest become ground memories, particularly when responded to with violence, as in the 1977 Labor Day Massacre as well as the Taksim Square protests of 2013. There is no apolitical public space. Space is political, filled with past protests and daily occupations. To the least imaginative (i.e. Erdogan), it is empty – to be filled with monuments to the present until a collective dissent reasserts the memories that resist erasure. Protest reclaims a "living heritage" rooted in the space.[32]

Erdogan used the police to crack down on the protests, thus escalating the scale of dissent. Protesters turned to social media (until the internet was blacked out), clandestine gatherings and night-time tagging of walls. Dialogue with space can sometimes be literal. Park occupiers inscribed the language of protest onto the landscape using spray paint. It was an effort to dialogue with power and with the space that met with resistance. Markings on surrounding walls, protesting the police presence in often humorous ways, resulted in daily maintenance of the park … mornings spent painting over graffiti in dull colors of yellow and gray.[33] Erdogan and the government criminalized graffiti, as well as the photographing and publication of graffiti. The Turkish term "çapulcu," translated as "hooligan" or "looter," was applied to graffiti artists, but eventually appropriated by the protesters as an identifier of resistance. Under a smiling picture of Erdogan, they sprayed "Everyday I am çapuling!" on the walls.[34]

Usually, landscape dialogue is not so direct as spraying words on walls. When the people rise up in disagreement, media, politicians and the protesters themselves magnify key words that include challenges to existing conditions and demands for change. These words of dialogue have power. The centrality of the park heightens their power and their tension. The space itself, already a rallying point, becomes a rallying cry … Tianamen, Tahrir or Taksim. Of course, not all public spaces experience dramatic dissent. Yet, protest dialogue, such as in Taksim Square, is not limited to the most prominent spaces. A common phrase during the protest in Instanbul – "Her Yer Taksim Her Yer Direniş" ("Everywhere is Taksim, everywhere is resistance") – suggests that any evaluation of a town square elsewhere in Turkey links, both spiritually and politically, with Taksim Square.[35] Each village gathering place or central square of national significance contains the potential for memories to re-emerge.

> What we have termed "ground memories" cannot be systematized, rendered coherent or smooth, fixed or frozen. They are hard to eliminate and tend to re-emerge upon any attempt to do so. Their availability as resources for resistance mean that ideological plays to expunge identity markers and, ultimately, protesting bodies from place are fundamentally problematic: multiple re-colonization acts ensue.[36]

It would be difficult to find a space free of place trauma, where people have not been oppressed or displaced or ignored. The challenge is not in finding these stories (as long as one is looking) but to then incorporate ground memories into the spatial evaluation process and eventual design.

7.11 Graffiti and other markings near Taksim Square at time of protests.

Dialogic methods use the body as a diagnostic tool. By statically placing ourselves in an unexpected position in the landscape, we experience conflict. By moving against traffic, we viscerally learn about norms of movement that shape the landscape. Through transgression of rules of conduct in public space, we examine the social structure of space by experiencing the effects of the rules.

1. Map trespass

Trespass can be method. The deliberate traverse of private property or public space managed as private confronts the division of space into parcels, single-purpose land use and the policing of those spaces. Trespass allows the designer to experience inaccessible places, particularly lightly populated areas of the city that have become derelict. Bradley Garrett describes a bodily engagement with the imposed structures of private place as "urban exploration."

> Urban exploration is a practice of researching, rediscovering and physically exploring temporary, obsolete, abandoned, derelict and infrastructural areas within built environments without permission to do so … Urban exploration can be connected to earlier forms of critical spatial engagement, the movement also speaks to the current political moment in unique ways. Urban explorers are one of many groups reacting to increased surveillance and control over urban space, playfully probing boundaries and weaknesses in urban security in a search for bizarre, beautiful and unregulated areas where they can build personal relationships to places.[37]

To explore the city, the designer places one's body in a place of unbelonging. This is bodily dissent. This will be a different experience for different bodies, whether a black woman, a person with a disability, a white man. Landowners, security and police respond in different ways to different bodies. The physical imposition of that which is "out of place" onto a place of privacy or singular use results in a revealing of the rules guiding that place. Think of a homeless person with all of their personal belongings occupying a plaza, asked to move along by the police, thus revealing the private nature of what was presumably public.

Of course, trespassing is not legal. While it is not acceptable to move into anyone's private space, many semi-public, abandoned and vacant spaces can be accessed without exposing oneself or others to harm. The designer can assess the pros and cons of flouting conventions or laws in the context of local conditions, social structures and surveillance. In the absence of actual trespass, the following mapping exercises can reveal the boundaries and social norms of a place:

- Map walls, fences and boundaries to measure an area's divisiveness … a potential proxy for fear and anti-social thinking. The degree of welcome-ness based on hardened boundaries will be different for each land use – industrial, commercial, residential – but within each land use, welcome-ness will vary, potentially pointing to other challenges.
- In a reverse of mapping enclosed fenced areas, map those open spaces that are publicly accessible (see Figure 7.12). This is not the same

American River Parkway

Cesar Chavez
Park

///. - Unfenced, accessible public space in the River District of Sacramento

7.12 Map of publicly accessible open space in the River District, Sacramento, California, most of which is associated with the rivers.

exercise as turning on a city's GIS layer of parks and open space, as many of the parks listed by the city may not be publicly accessible. The ratio of open space (that is truly open) to overall urban space serves as a welcome-ness indicator. Like the first mapping exercise, the map should cover an area larger than the site, likely a neighborhood.

- Map cameras and surveillance in urban areas, another proxy for the degree of welcome-ness or public-ness of a place.

Interview people experiencing homelessness. No one in the urban environment is as aware of the obvious and not so obvious boundaries as unhoused people. Their experience of inhabiting streets and parks and stoops attunes them to the particular spatial "mappings" others impose on the city, so that they come to know where they will not be bothered by police or thieves. These places correlate with softer boundaries that allow for trespass while maintaining a certain amount of temporary privacy. Practice sensitivity in approaching and talking to people living on the streets.[38] It is usually very clear if someone is willing to talk or not. Do not ask directly about trespassing. Potential questions include:

- Where do you go when you first wake up? How do you get there?
- Where are your favorite places?
- Which places do you avoid? And why?

Talk to the police. Police provide invaluable information around their priorities and "public" opinion. In a study of homeless transit use, I interviewed a police captain for the Sacramento Transit Police who described the process of maintaining transit safety and discouraging fare evasion. Previously isolated light rail stops (in my mind) became linked with each other as part of a system of surveillance and patrol. I also learned who the transit police were maintaining the system for (… not unhoused people).

2. Protest

Join a protest to better understand dissent. March to the state capitol. Hold a sign at the local farmer's market. If trespass is an individual's dissent of existing systems of property, then protest is a group's dissent of similar systems. It is collective action. To that end, protest is more relational; there will be negotiations with fellow protesters. Talk to them. We do not even have to agree with the protest's main ideas or aims.

Some of the most revealing accounts of the January 6 storming of the United States Capitol building came from journalists walking with the marchers to the Capitol (then presumably stopping during the assault on the building).

Each protest reveals something different about space. Take Back the Night, a series of annual marches by women to call attention to sexual and domestic violence, asks the question: Why do women need to be in a large group to feel safe at night? For designers, the question inspires an examination of public space in relation to safety, lighting and gatherings. It brings a temporal element to the process and encourages us to dialogue with space at night as well as the day. Other protest movements speak specifically to the configuration of public space. The Critical Mass bike rides that happened in San Francisco challenged the single-purpose nature of streets. Large groups of cyclists would overwhelm streets devoted to cars, calling for the city to attend to their spatial and economic needs.

To participate in a protest could also include marking the spatial patterns and movements of past protests. I did not personally participate in the march for voting rights at the Edmund Pettus Bridge in 1965, but I have visited with a civil rights historian to examine the spaces of Selma, the bridge and the larger state of Alabama.

It is not enough to read about dissent or visit a site. Landscape dialogue is not just a listening; it is speaking. The purpose of a praxis dedicated to participation in, or study of, dissent is to inspire the design process to move towards more equitable and sustainable design. To that end, the following additional ideas can transform protest into understanding and advocacy:

1. Plot a historic timeline of power and events that influence and take place within a landscape. Illustrate power in terms of *who* controls the landscape. In contrast, portray events as initiated by people, particularly those who resisted and pushed back against those in power. (See retrospective methods in Chapter 4 of Stahlschmidt et al., *Landscape Analysis*).

2. Diagram current power dynamics related to the landscape in question (see Figure 7.13). In contrast to the relational diagrams of Chapter 5, explore the power dynamics of events by showing historical images and hierarchies. Who owns the space? A property or parcel map remains a critical component of dialogue with the landscape and the people who

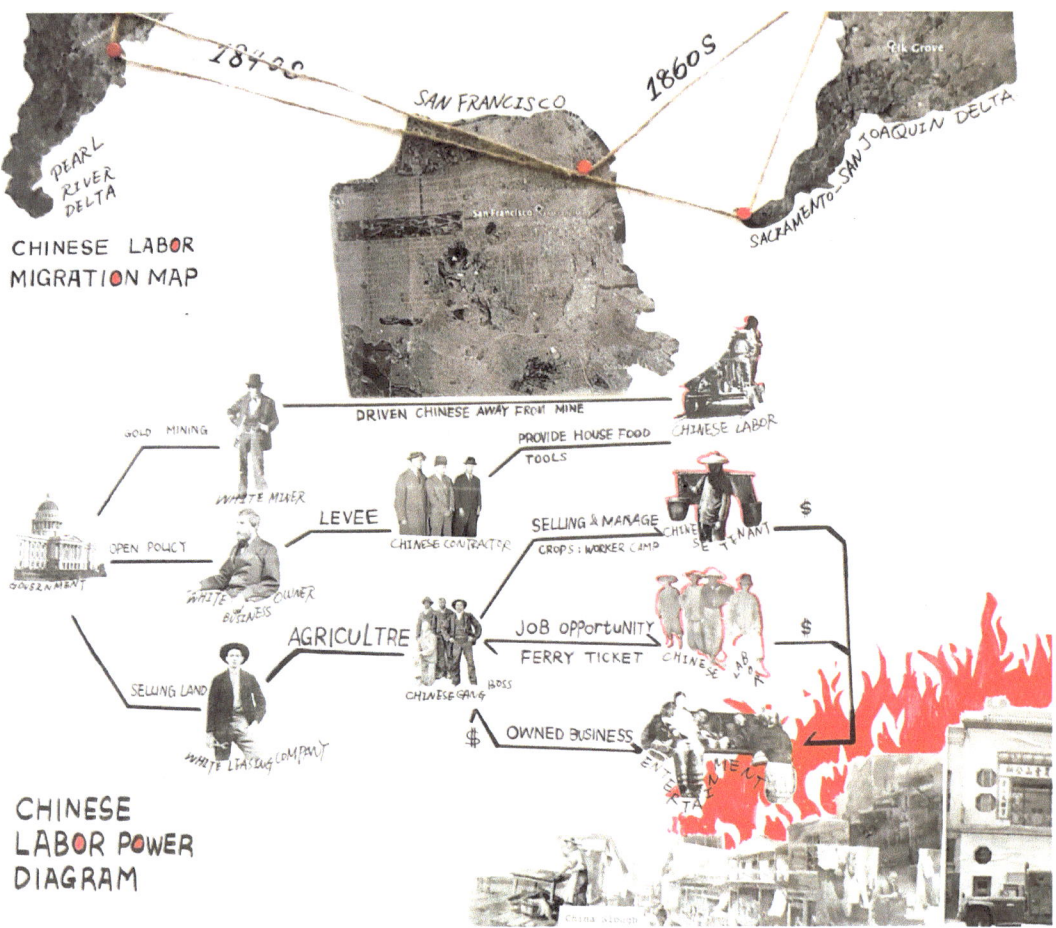

7.13 Diagram of power dynamics of people and events having to do with Chinese labor moving from China to San Francisco to the Central Valley of California.

own it. Who controls the space? This question relates to the maintenance and accessibility of a landscape, exploring direct and indirect barriers to inhabiting public space. How are the boundaries of the space defined and maintained? The edges of a space as experienced by people, particularly the entry into a space, reflects ownership, control and maintenance.

3. For a more visceral experience of landscape dialogue, participate in the protest event itself. Track how people move through and within a march. Take snapshots every block. What parts of the street do people choose to occupy, to move through, to rest in? How is traffic interrupted? Where do people assemble and what does it look like? Then take that and make design claims regarding the street. What makes for a good protest street?

NOTES

1. Juan Williams and Julian Bond, *Eyes on the Prize: America's Civil Rights Years, 1954–1965*, 25th anniversary edition (New York, NY: Penguin Books, 2013), 273.

2. "Global Slavery Index," https://www.globalslaveryindex.org/.

3. Taylor Branch, *Parting the Waters: America in the King Years 1954–63* (New York, NY: Simon & Schuster, 2007).

4. Jason Sokol, *There Goes My Everything: White Southerners in the Age of Civil Rights, 1945–1975* (New York, NY: Knopf Doubleday Publishing Group, 2008).

5. Sokol, *There Goes My Everything*, 69.

6. Sokol, *There Goes My Everything*.

7. Nicholas A. Brown and Sarah E. Kanouse, *Re-Collecting Black Hawk: Landscape, Memory, and Power in the American Midwest* (Pittsburgh, PA: University of Pittsburgh Press, 2015).

8. Samara Bonsey and Mandy Deitering, "1996 All OSU Boycott & March," *Histories of Students of Color at Oregon State University*, January 2016, http://photohistory.oregonstate.edu/works/untold-stories-guide/1996-all-osu-boycott–march.

9. I should note that the march took place against specific local laws, sans permit.

10. John Lewis, *Across That Bridge: A Vision for Change and the Future of America*, Reprint edition (New York, NY: Legacy Lit, 2017).

11. An argument could be made that the protesters did not need a permit, as freedom of speech and assembly are rights guaranteed by the first amendment of the Constitution, while movement within and across state boundaries in the United States is also a right (*Corfield v. Coryell*, 6 Fed. Cas. 546 [1823]). See also Article 13 of the United Nations, "The Universal Declaration of Human Rights," 1948 and Hagar Kotef, *Movement and the Ordering of Freedom: On Liberal Governances of Mobility* (Durham, NC: Duke University Press Books, 2015). However, freedom of speech can be limited by requiring a permit for gatherings, according to the "time, place and manner" restriction, if it is found "the manner of expression is basically incompatible with the normal activity of a particular place at a particular time" *Grayned v. City of Rockford* (1972).

12. Don Mitchell and Lynn A. Staeheli, "Permitting Protest: Parsing the Fine Geography of Dissent in America," *International Journal of Urban and Regional Research* 29, no. 4 (2005): 796–813, https://doi.org/10.1111/j.1468-2427.2005.00622.x.

13. Mark Purcell, "Resisting Neoliberalization: Communicative Planning or Counter-Hegemonic Movements?" *Planning Theory* 8, no. 2 (May 1, 2009): 140–165, https://doi.org/10.1177/1473095209102232.

14. Kenneth Foote, "On the Edge of Memory: Uneasy Legacies of Dissent, Terror, and Violence in the American Landscape," *Social Science Quarterly* 97, no. 1 (2016): 115–122, https://doi.org/10.1111/ssqu.12259.

15. Don Mitchell, "New Axioms for Reading the Landscape: Paying Attention to Political Economy and Social Justice," in *Political Economies of Landscape Change: Places of Integrative Power*, ed. James L. Wescoat and Douglas M. Johnston (Dordrecht: Springer Netherlands, 2008), 29–50, https://doi.org/10.1007/978-1-4020-5849-3_2.

16. Pierre Bourdieu, *Distinction: A Social Critique of the Judgement of Taste* (Cambridge, MA: Harvard University Press, 1984); Michel Foucault, *Discipline & Punish: The Birth of the Prison*, translated by Alan Sheridan, 2nd edition (New York, NY: Vintage Books, 1995).

17. Andrew Hussey, *The Beast at Heaven's Gate: Georges Bataille and the Art of Transgression* (Amsterdam: Rodopi, 2006).

18. David Pinder, "Transgression," in *The Dictionary of Human Geography*, ed. Derek Gregory, Ron Johnston, Geraldine Pratt, Michael Watts and Sarah Whatmore (Chichester: John Wiley & Sons, 2011).

19. Chris Jenks, *Transgression* (London: Routledge 2003), 2.

20. Tim Cresswell, *In Place/Out of Place: Geography, Ideology, and Transgression* (Minneapolis, MN: University of Minnesota Press, 1996).

21. Jenks, *Transgression*, 7.

22. Tim Cresswell, "Falling Down: Resistance as Diagnostic," in *Entanglements of Power: Geographies of Domination/Resistance*, ed. Ronan Paddison, C. Philo, P. Routledge and J. Sharp (London: Routledge, 2000), 256–268.

23. Michel de Certeau, *The Practice of Everyday Life*, trans. Steven Rendall, 2nd edition (1984; repr., Encinitas, CA: University of California Press, 2002).

24. Deborah Reed-Danahay, "Talking about Resistance: Ethnography and Theory in Rural France," *Anthropological Quarterly* 66, no. 4 (1993): 221–229, https://doi.org/10.2307/3318065.

25. Michael K. Middleton, "Housing, Not Handcuffs: Homeless Misrecognition and 'SafeGround Sacramento's' Homeless Activism," *Communication, Culture & Critique* 7, no. 3 (September 1, 2014): 320–337, https://doi.org/10.1111/cccr.12055.

26. Kotef, *Movement and the Ordering of Freedom*.

27. See Chapter 2 of John Tomlinson, *The Culture of Speed: The Coming of Immediacy* (London: SAGE, 2007).

28. Tomlinson, *The Culture of Speed*, 36.

29. Murat Gül, John Dee, and Cahide Nur Cünük, "Istanbul's Taksim Square and Gezi Park: The Place of Protest and the Ideology of Place," *Journal of Architecture and Urbanism* 38, no. 1 (January 2, 2014): 63–72, https://doi.org/10.3846/20297955.2014.902185.

30. Christopher Whitehead and Gönül Bozoğlu, "Protest, Bodies, and the Grounds of Memory: Taksim Square as 'Heritage Site' and the 2013 Gezi Protests," *Heritage & Society* 9, no. 2 (July 2, 2016): 111–136, https://doi.org/10.1080/2159032X.2017.1301084.

31. Whitehead and Bozoğlu, "Protest, Bodies, and the Grounds of Memory," 115.

32. Whitehead and Bozoğlu, "Protest, Bodies, and the Grounds of Memory," 116.

33. Kyle T. Evered, "Erasing the Place of Dissent: Inscriptions and Eliminations of Gezi Park Graffiti," *Area* 51, no. 1 (2019): 155–165, https://doi.org/10.1111/area.12439.

34. Evered, "Erasing the Place of Dissent," 158.

35. Whitehead and Bozoğlu, "Protest, Bodies, and the Grounds of Memory."

36. Whitehead and Bozoğlu, "Protest, Bodies, and the Grounds of Memory," 130.

37. Bradley L. Garrett, "Undertaking Recreational Trespass: Urban Exploration and Infiltration," *Transactions of the Institute of British Geographers* 39, no. 1 (January 1, 2014): 1, https://doi.org/10.1111/tran.12001.

38. Federal guidelines on social research in the Unites States do not specifically identify adult people who are unhoused as a vulnerable population (National Research Act, 1974). However, in my research with this population I have encountered people who met many of the criteria for vulnerability. They may be experiencing a mental illness, threats from other people or extreme deprivations which necessitate transparency and compassion. If approaching an unhoused person, first introduce yourself. Explain your purpose in evaluating public space. Listen before talking. And constantly assess their body language for signs of discomfort which may necessitate curtailing the interview. That being said, I have found most unhoused people happy to talk about their day. Read the article "Ethics, Reflexivity and Research: Encounters with Homeless People" by Paul Cloke, Phil Cook, Jerry Cursons, Paul Millbourne and Rebekah Widdowfield (*Ethics, Place and Environment* 3, no. 2 (2000): 133–144) for a discussion.

CHAPTER 8

Landscape formation: openness to personal change

The landscape shapes the designer as the designer shapes the landscape. I return to Hawaii to focus on the impact of the landscape on the designer.

THE WINDWARD SIDE

At the beginning of the Hawaii Sustainable Roads project, the design team sat in a rented house in Kailua on the windward side, listening to a native Hawaiian facilitator talk about her Hawaii. We were all anxious to go out to the highway and begin analysis but she spent time unfurling stories of the island. At a break in the meeting, she asked me to identify what emotion I was feeling. I momentarily thought, in this professional setting, what kind of question is that? I was relaxed. We had had a wonderful Hawaiian meal the night before. There was a peaceful breeze coming through the open windows … but "relaxed" is not really an emotion. What was I supposed to say? That I was confused, that I was impatient to get out to the landscape? So I said "curious," which was the truth. She responded: "That is a good place to be." Break over.

She facilitated another exercise. As outsiders (which most of us were), we negotiated with an insider about our place in this space. We discussed the west side of the island (the dry/leeward side) and the native movement to re-establish a sovereign Hawaii led by Hawaiians. We discussed the east side of the island (the wet/windward side) and tourism and the *ahupua'a*. Gradually, we let go of our ideas about the landscape, about the highway, and grabbed a hold of the idea of a people. Of the people of Hawaii, this strange mixture of native Hawaiians, Japanese Americans and *haole* descending in planes to crawl over the beaches. It was about

DOI: 10.4324/9781003158943-9

8.1 *Kuapa* (rock walls) surrounding a former fishpond on the east coast of Oahu.

the highway; it was about the landscape. But the design did not matter unless we first understood the people, their traditions, habits and desires. Sustainability of the environment would not work without sustainability of the community.

When we eventually arrived at the highway as a group of experts, we made no mention of sustainability. We did not even look at the highway. A historian with a deep knowledge of the Island of Oahu took us to a fishpond at the mouth of the Kahana Valley, where we examined the *kuapu* (rock walls) that filtered the sea water moving into and out of the pond, providing habitat unique to much of Polynesia. He spoke of the *ahi'i* (chiefs) who ruled the *ahupua'a* before Kamehameha conquered the island. The people lived within each watershed, harvesting from the head of the valley, growing taro on every flat area and harvesting fish in over 100 fishponds along the coast. A place of abundance. When tourist and second homes landed on the islands en masse, this windward side transitioned to a community of second homes, hotels and gas stations. The island can support a richness of life, a richness of cultures. The sustainability expert and the landscape architect from a major transportation agency began to question universal standards of sustainability in the face of the uniqueness of each unfolding valley. A one-size-fits-all site analysis misses the heterogeneous nature of place.

THE LEEWARD SIDE

On the other side of Oahu, a smaller group of us drive west past the Disney resorts, shoot through a narrow road between cliff and sea, before the land opens up, invasive grasses draping the rocky slopes up the Nanakuli Valley and then Waianae itself. This is the dry side, in island-parlance, the leeward side. Without the rains of the windward side of Oahu, the Waianae area lacks tropical lushness and, thus, tourists.

Lunchtime, we park at the Tamura Supermarket, grab sandwiches and drive a few blocks south to one of the beaches. It's a weekday and it's windy. Very few people are here. Leaving our orange highway vests in the car, we walk out onto the sand and a tanned man with long hair walking along the beach veers toward us. "Hey, you guys need to be careful! This isn't a place for *haoles*. Try back at the resorts," he says, gesturing to the south.

Waianae has the highest concentration of native Hawaiians on Oahu. A social life centered around the supermarket and a web of familial relations extending across the island and the Pacific. With the exception of a prime surfing area near the end of the road at Mākaha Beach Park, it lacks resorts, only beach, highway and then one-story residences surrounded by painted concrete block walls. The beaches seem more like community parks with long rows of picnic tables and large family groups barbecuing, even on a weekday, even in this wind. Three kids attempt to set up a badminton net, while their younger siblings go back to their parents wanting to be included in the stories and laughter.

After lunch, we return to the highway, a critical link to Honolulu and goods and services from the mainland for this community. Before the arrival of Cook in the eighteenth century, Waianae never supported a large concentration of Hawaiians. Now, the small communities have entrenched themselves into the fluctuating spaces between the expansive beaches and the hillsides. They are suspicious of outsiders, of development and actively work toward the restoration of a native Hawaii. Elsewhere, on Oahu, the visitor can ignore the history of violence bound up in the landscape, but here in Waianae, it seems more present, the dryness of the slopes echoing the paucity of resources. It is this public road, this tenuous connection at times, that now provides them food, books and badminton

8.2 Map of the highway along the leeward side of Hawaii and the Wai'anae moku.

sets. They can no longer sustain themselves with the produce of their place; the *ahupua'a* survives but is breached, colonized, but entrenched in the landscape and memory. The highway continuously probes into the spaces of everyday lives. Pavement finds us. It may narrow in approach as the tourists dissipate but its ubiquity ensures everyone connects.

It is this road we are analyzing to learn how to make it more sustainable. This road connecting the residents of Waianae to their brother-in-law in California, their aunt in Tokyo, or their friends on the windward side. This road that brings surfers from Australia. This road which shuts down during increasingly frequent storm events. It is a road between land and sea: a tenuous stream of asphalt.

After a few days of driving around, despite my expertise in highway design and landscape, I had to admit that we would not understand Oahu as others might understand it. As it was necessary to understand it. "The secrets of the land die with the people of the land."[1] We might understand cars, speed and infrastructure. We might understand landscape processes of stormwater runoff and the encroachment of invasive species. We might even understand the series of linked *ahupua'a*. But we would not understand the people and their passions, the way they related to the

8.3 Hawaii State Highway 93 on the leeward side running south through *ahupua'a*.

place, the highway, the way they raged against the "American behemoth."[2] That would take much longer. It would take dialogue. In the midst of the babble of heteronomous Hawaiian voices, I sat in a car, trapped in a space or position from which I attempted to engage with others, interpret their meanings and use that to inform design. Bakhtin suggests that complete understanding of another is impossible, due to the different meanings present even when using the same words.[3] We sit before another as separate individuals and never fully reconcile the meanings of the words we hear, spoken amidst an already existing difference.[4]

There is hope. While we cannot understand another (or the landscape) completely (a condition which would eliminate the need for dialogue), we can reach a mutually contingent understanding of situations and places. This understanding might be limited by time and location and cultures, but it does crop up and may yield new insights, insights that could not come from the already entrenched.

There exists a very strong, but one-sided and thus untrustworthy idea that in order to better understand a foreign culture, one must enter into it, forgetting one's own, and view the world through the eyes of this foreign culture. This idea, as I said, is one-sided. Of course, a certain entry as a living being into a foreign culture, the possibility of seeing the world through its eyes is a necessary part of the process of understanding it; but if this were the only aspect of this understanding it would merely be duplication and would not entail anything new or enriching. Creative understanding does not renounce itself, its own place in time, or its own culture; and it forgets nothing.[5]

I did not believe then, nor do I believe now, that I had nothing to offer to the design of Hawaii's highways through the process of landscape dialogue, but the creative understanding necessary to impact the landscape never got off the ground. Our four days on Oahu driving the highways were negligible. We could not experience in any way the pain of conquest from the eyes of the conquered. We could not overcome cross-cultural barriers to respond to ideological ideas of "island" and "sovereignty." Conversely, we had little time to engage with Honolulu's entrenched engineering bureaucracy with enormous stakes in perpetuating an infrastructure of pavement and speed. The lack of interest in engagement from the engineers opened my eyes to the absurdity of applying national AASHTO standards to a tropical island. Hawaii had roads but they were not Hawaiian roads. The epiphany has since shaped my advocacy for people over cars in the landscape, proximate materials over petroleum products.

The Hawaiian landscape, its network of entangled relations undergirded by conquest, changed me. I returned to the mainland thinking about its power geometries. Its dissent hidden behind the veil of tourism. I acknowledged my own privilege as an educated designer – a privilege cultivated within a system of contracts, drawings and social agreements of the way things are designed. A privilege that can itself obscure my perception of people–landscape relations. At the same time, a privilege bounded by a narrow frame of roadside improvements, plants and wayfinding for which my position as landscape architect has trained me, and as assumed by others, has *not* trained me for evaluating other aspects of the landscape like the road itself. It is a positional privilege as long as I operate within social and legal constraints. This privilege contrasts with the residents who are assumed to lack landscape "expertise."

Ku'ia kahele aka na'au ha'aha'a [A humble person walks carefully so as not to hurt others].

8.4 Power relations in the Oahu landscape, west side of the island.

In my experience as a designer, humility is rare. It as if the Designer with a capital "D" burst from the womb fully formed – a paragon of mistake-free, heroic gesture. I have been involved in projects in which I articulate a mistake, take responsibility and suggest how to correct it, only to notice how no one else admits to error or failure. (I cannot be the only one making mistakes here!) An inability to admit mistakes suggests insecurity, and insecurity is death to creative design. Many designers feel they must portray a "put-togetherness" that is simply alien to the human condition and certainly an anathema to understanding the landscape. We often do not have answers. We may have no idea how to get them. Yet, our clients do not want to hear about limits or mistakes. Learning is stilled. For the project in Hawaii, the very premise of creating more sustainable highways was flawed, gathering information about a place would never make up for the intrinsic, unsustainable nature of highways with their petroleum asphalt and individualistic transport.

In landscape dialogue, the designer is open to *personal* change. A dialogue with the landscape is an exchange, a negotiation of interdependent positions from different positions. Landscape dialogue learns from the whole place as "convivial engagement," while being responsive to new ways of thinking about the landscape, particularly from marginalized positions. The designer practices ethnography, a deep dialogue with the Other.[6] How does our own subjectivity influence/interact with another person? With the landscape? What positive and negative qualities of the landscape can inspire personal change? Designers experience beauty and abundance, distressed systems and inequities. Collectively, these experiences transform our relationship with the landscape. It is a reflective way of perceiving the landscape. Reflectivity acknowledges the lens through which the designer sees the world as deeply personal/cultural.

Landscape dialogue, whether positive or negative, is *formational*. The designer immersed in place changes through the myriad of influences and inputs that stem from being open to place. This chapter asks: what would it look like to rely on landscape as a formative part of one's life? The power of landscape – not a power bequeathed from planners or politicians but power inherent in the movement of wind, air, people and materials as they continuously shape a place – inspires us to be more sensitive and stronger designers, more passionate and thoughtful people.

REFLEXIVITY

The purpose of landscape dialogue is to understand the landscape *and* to understand oneself. In fact, it could be said that an understanding of oneself, however partial, is critical to understanding the landscape. We now turn to the self – the subject walking through the garden – and the person and position of perceiving the landscape we examine.

Like many professions, the pursuit of design excellence exists as if separate from the personal experience and identity of the designer. A possible exception is the Star-chitect who becomes identified in a public way with their project or style (but not necessarily their humanity). For most designers, we may experience growth throughout our careers, learning new skills, shedding old notions ... constantly becoming, but suppressing in many ways our personal foibles and ignorance. In this context, we learn and grow through, within and because of the landscape. To share that growth, what we learn, might reveal a personal vulnerability, the assumption that we have made past errors. Aspects of the current design profession do not encourage personal reflection. In marketing one's design prowess, the process of responding to public and private clients requesting qualifications or a proposal, designers must transform (on the surface) into people who can do no wrong. There is no room for imperfection.

Personal reflection is critical to dialogue with the landscape. We must know from where we listen and from where we speak. In the social sciences, feminist researchers question positivist notions of Western science that claim the landscape can be perceived from "nowhere" as an objective observer.[7] The researcher is always positioned somewhere. The landscape is one vast subject of research, related to geography, planning, urban design, landscape architecture and anthropology, and as such, can be approached through myriad positions, both in terms of discipline and worldview and in terms of a spatial location with a physical arrival, sequence and exit. This dual concept of physical located-ness and philosophical point of view is *positionality*. Acknowledging positionality forces the researcher to consider one's limits and constraints that affect research. Our analysis is informed by who we are and how we see.

For designers, positionality can be difficult. Most designers accumulate years of education, along with undergoing some sort of licensing process to ensure the designer knows how to design. We occupy a position of privilege in terms of our education, our language and our sensibilities. Education can certainly be a positive experience of personal and

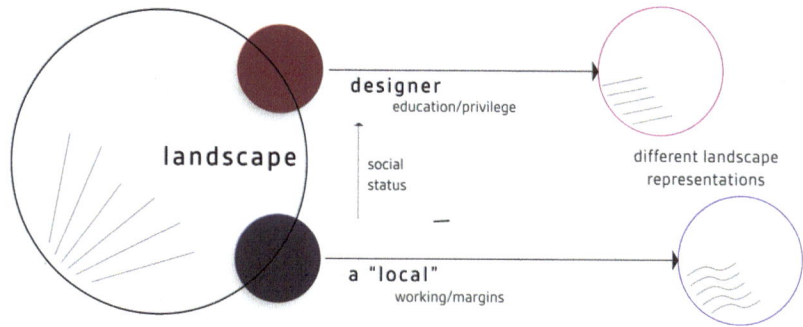

a)

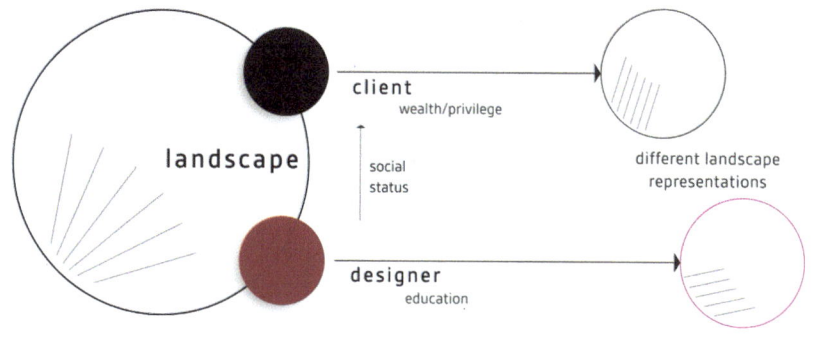

b)

8.5 Positionality filters people's view or ideas of the landscape: (a) designer as "higher" in social status than the local resident; (b) designer as "lower" in social status in comparison with wealthy clients.

professional growth but it can also separate us from people without an education; a lack of education may intersect with class, race and gender. (The process of education and training itself may be experienced differently by women and minorities, suggesting a need for greater diversity in the design fields.)[8] Thus, designers can separate from others *through the process of becoming a designer.* We must consciously practice designing for the interests of others, embracing difference, a practice that starts with understanding one's own positionality.

An understanding of self – our position, privilege and personality – changes our approach to the landscape. Our approach affects what we learn about a place. This means we need to acknowledge our positionality and examine its influence on our landscape dialogue. The practice of acknowledging and understanding our positionality is called *reflexivity.* It takes thoughtful consideration of background, gender, ethnicity and amount of education, as well as careful discussions of identity with other people. We do this constantly, aware of our multiple roles in the landscape as teacher, student, friend, visitor, resident, professional and/or laborer.

Katz situates the researcher (and by extension, the designer) in a place of "between-ness" – not quite the subject of dialogue but also not the speaker.[9] Designers position themselves within the relational diagrams of the landscape (see Chapter 4), within networks of power, as a bridge between landscape and people.

As discussed in Chapter 2, site analysis is a process of instrumental rationality – "professional activity consists of instrumental problem solving made rigorous by the application of scientific theory and technique."[10] In instrumental rationality, there is no need for reflection, because universal principles of technical know-how are applied. These principles are supposedly disconnected from the person analyzing the landscape. Positionality is assumed (to be objective). Without reflection, however, the designer loses the ability to empathize, imagine and, ultimately, fit a design into a local context.

Reflexivity is already something each designer practices. Donald Schon, in an early study of how professionals learn as they practice, contrasts instrumental rationality with the "reflective practitioner."[11] He found that professionals do not draw from their academic learning as much as the practice of an ongoing, qualitative self-reflection that gives feedback on how things are developing. In Schon's findings, professionals, including designers, did not undertake a series of pre-determined steps to generate analysis and design. Rather, for each project, they practiced "reflection in action." The professional might use the language of instrumental rationality regarding their profession but could not explain their actual practice – why they do certain things and not others – in technical terms. Practitioners often act without necessarily being able to describe the reasons for their actions (beforehand). They make judgements, quick qualitative decisions.

Knowledge then comes through doing. While acting, the practitioner may question their actions – reflecting on how it is qualitatively proceeding – to make sense of a phenomenon. Defining those phenomena and their challenges frames the problems to be addressed:

When we set the problem, we select what we will treat as the "things" of the situation, we set the boundaries of our attention to it, and we impose upon it a coherence which allows us to say what is wrong and in what directions the situation needs to be changed. Problem setting is a process in which, interactively, we name the things to which we will attend and frame the context in which we will attend to them.[12]

We think about the landscape as we move through it. As we sketch it or experience it. As we "analyze" it. Not as a separate thing. The reflection becomes part of the action – leading to improvisation and adaptation.

Now, feminist reflexivity and Schon's reflection-in-action are not incompatible. A better understanding of our own positionality means a greater ability to reflect in action, particularly in "letting the imagination go visiting" of specific situations in time and space.[13] In Schon's reflection in action, we, as professionals, can foster a more attentive practice that embraces intuitive and contextual thought applicable to the specific situation, even if this practice is difficult to articulate.

This is not a solo process, despite reflexivity's focus on self. It is a dialogic process. It happens within spaces that slowly reveal a network of relations, experiencing the same landscape within the same time frame. We do not exist as separate individuals but constantly dialogue with the landscape and others, a dialogue which shapes who we are as designers. The geographer Gillian Rose suggests reflexivity should be transparent ... that is, the researcher should communicate their point of view during and after research.[14] The same applies to designers. We make known our positionality as designers to the client, to the team and to the public, negotiating our privilege and expertise with others. Others can then evaluate our approach to the landscape, our findings and our representations of it, incorporating this perspective into their own analysis/dialogue. Reflexivity then is a dialogue aware of our own position and a communication of that position.

PERSONAL FORMATION

The theologian Robert Mulholland distinguishes between two types of reading text: the informational and the formational.[15] In the informational reading, the reader selects a text to be read, then extracts information to increase one's knowledge of the topic (see Chapter 2). "We read the text analytically, viewing it as an object over which we as subject exercise our control, to ensure that it conforms more or less comfortably to our purposes."[16] This informational approach is how we frequently "read" the landscape. It is the approach of instrumental rationality. In contrast, in *formational* reading the reader sits before the text in an open and engaged manner, plumbing its depths so that the depths of the self might also be plumbed. Formational reading gives power to the text to shape the reader.

Schwager summarized Mullholland's formational approach to reading in a series of couplets:[17]

Informational reading	Formational reading
A linear process	An in-depth process
Seeks to master the text	Allows the text to master us
The text as an object to use	The text as a subject that shapes us
Analytical, critical and judgmental approach	Humble, detached, willing, loving approach
Problem-solving mentality	Openness to mystery

If we are going to read the landscape, we need to be ready to respond to what the landscape is telling us and, in that response, allow the landscape to shape us. The landscape has power – a power to move and work into our lives. To inspire us. To tell us things, some of which might be uncomfortable to hear. It is the formational idea of reading the landscape, in which the landscape is reading us, in a sense, that binds the designer to place.

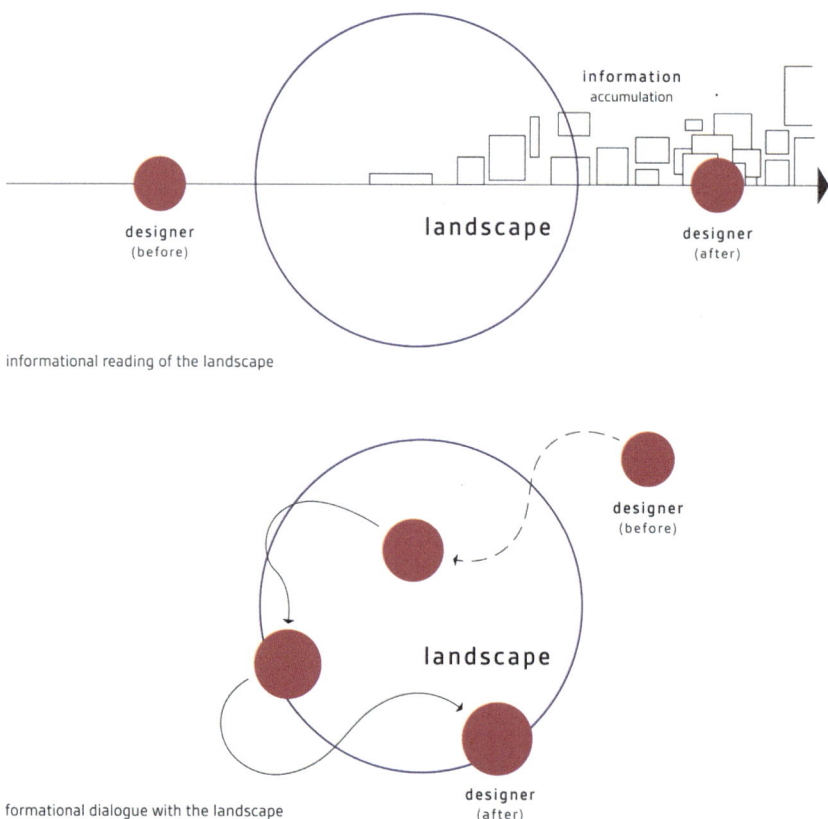

8.6 Diagram of reading in the landscape formationally, leading to personal growth.

In his novel, *Medicine Walk*, Richard Wagamese (Ojibwe) wrestles with the taut relationship between Indian and landscape.[18] Franklin, the youthful protagonist who has been raised by the "old man" to honor the forest for its providence, responds to his biological father's call to visit him as he is dying. He hasn't seen his father in years. His father is part Ojibwe, part Scot, working in a lumber town, always working with, and yet away from, the land. Franklin goes to see him and takes him out of his squalid townhome and into the wilderness to bury him as a Native American warrior, facing the eastern sun … this father who he has rarely met. Along the way, Franklin tells him of some cliff drawings:

> "There's signs up there. Symbols. Painted right into the rock. When the old man took me there the first time he said it was sacred because no one can ever figure out how come the paintings never faded. They been there a powerful long time."
>
> "I hearda places like that. Never been to them. Never seen them."
>
> "Seems like you should see it now."
>
> "We gotta climb to it."
>
> "The horse can get you most of the way. I'll lug you the best I can up the rest."
>
> "Sounds like a lotta work for a few paintings."
>
> "Lotta times, I guess, you never know what you need until you lay eyes on it."
>
> "You gotta be a philosopher," his father said.
>
> The kid looked at him and shook his head. "Not so much. I mean, out here things just come all on their own sometimes. Thoughts, ideas, stuff I never really had a head for."[19]

After some whiskey, the father agrees to go. Franklin helps him up onto the horse and walks alongside higher up into the mountains. It gets too steep and narrow, so they abandon the horse, and Franklin helps his sick father clamber up, eventually easing his father out onto a ledge below the cliff face.

> It took a moment for his breathing to settle and when he finally raised his eyes to look at the cliff face his mouth draped open.
>
> "Damn, Frank," he said.
>
> The kid sat down beside him and they both stared up at the wall of rock. There were symbols painted in a dull red, black, and a

stark greyish white. There were birds, oddly shaped animals, what appeared to be horses and bison, horned beings, stars, and assorted lines and shapes. The drawings stretched a full twenty feet up and covered the entire wall. They studied them without speaking for a long time.

"Take me up to it," his father said quietly.

The kid stood up and helped him stand. Together they shuffled to the face of the cliff. His father reached out and put his hand on the rock. Then he slid it over and covered a small dog-like shape and raised his head to look up at the array.

"What do they signify?" Eldon asked.

"I don't know. Near as I can figure they're stories. I reckon some are about traveling. That's how they feel to me. Others are about what someone seen in their life. The old man doesn't think anyone ever figured them out."

"Ain't a powerful lotta good if ya can't figure 'em out."

The kid shrugged. "I sorta think you gotta let a mystery be a mystery for it to give you anything. You ever learn any Indian stuff?"

His father lowered his gaze. He turned his back to the wall and slid down to sit. He brushed a hand over his forehead and closed his eyes to have a deep breath. "Nah," he said finally. "Most of the time I was just tryin' to survive. Belly fulla beans beats a head fulla thinkin'. Stories never seemed likely to keep a guy goin'. Savvy?"

"I guess," the kid said. "Me, I always wanted to know more about where I come from." The kid took out his makings and rolled them each a smoke. They lit up and smoked quietly for a minute or two. "I could come and sit here for hours. I spent three days here once when I was thirteen. Sorta thought if I spent enough time studying them drawings I could figure out what they were supposed to tell me."

"They ever?"

An eagle drifted over the valley. There was a yap of coyotes from somewhere below and the snap of a limb as something big moved through the trees above them. "Not really, I guess. Nothin' real, least ways," the kid said after awhile. "But it seemed to me no one came here no more. Like they forgot it was here. That made me sad. So I kept comin' so there'd at least be someone even if I didn't know how to read 'em or get what it was they were trying' to say. At least there was someone."[20]

At first, the father can only curse, astounded at the power of the drawings. But then, in his town-ness, his unawareness of the (native) landscape, he wants to know, to understand, to have them interpreted. And without that interpretation, they are "not much good." Then, the kid speaks in the language of the formational, letting a "mystery be a mystery," but still as a consumer … what can the drawings "give" to the reader. As the father contemplates his own Indian-ness, he continues to respond in an informational manner, interpreting his son's question about "Indian stuff" as a question about information, about "thinkin." After some quiet reflection, during which the animal world reasserts itself in the present, the kid confesses that he could never understand the drawings (informationally), but their transformational power is evident in the time he spent before them. He has moved from an interpreter of the drawings to a caretaker of the drawings, now giving back to them through acknowledgement and his consistent presence. He has changed.

During the Paynes Prairie interpretive center and trail network project, I found myself meandering along the edges of Paynes Praire State Park in central Florida. I walked with others on levees and boardwalks to observe prairie wildlife and overlook the sinks. Wildlife, such as snapping turtles, alligators, bison and herons, inhabited the prairie, attracted to its nutrient rich wetland vegetation. Trails skirt the edge because the center of the prairie contains marshy conditions impossible to traverse (and

8.7 Photo-simulation of proposed trailhead of Sweetwater Branch of Paynes Prairie.

8.8 Paynes Prairie with constructed boardwalk.

to provide wildlife safe spaces for living). As a designer, I was intrigued by the expansiveness of the landscape where the lack of topography generated mystery, as one could not get high enough to see into the center. The expansiveness opened up possibilities of design. The dual notion of the danger of the prairie (alligators, snakes, sinkholes) and its vulnerability (to drainage, chemical and nutrient input, predation) led to a certain sensitivity to the surface of the landscape. This consideration of expansiveness and sensitivity yielded both an outward-focused and inward-focused design – curving boardwalks into the heart of the wetlands and nodes at arc intersections for interpretation and contemplation. But beyond the design, it also led me, as the designer, to tackle bolder, more extensive geometries. (The proposed solar observation tower has not yet been built.) My design opened up to larger possibilities. I took chances. The dialogue with the prairie landscape changed my approach. It changed me, as designer.

Before taking responsibility for the landscape and one's designs, reflexivity suggests the designer should know themselves. It would be difficult to engage in meaningful dialogue without first understanding one's positionality and how one approaches life/design/landscape. To do that, a new designer should, even if economics and locational constraints

preclude re-visiting one's local place, engage with their home place, a process that can take a lifetime of dialogue.

Menna Agha, an architecture professor at Carleton University, spent years researching her home place and her Nubian kin through the lens of architecture. She was raised in Qustul, a village in Egypt of displaced peoples along the Nile River, displaced by the construction of the High Dam at Aswan which flooded their town in the 1960s. How did the buildings of Qustul get built? Who located them, who designed them, and who built them? Qustul was one of 42 displaced communities which the State called "New Nubia" but local Nubians called "Tahgeer" – the place of displacement and of pain and loss. Through a series of interviews, she learned that while her home was designed and constructed by men from the village, it was her grandmother who "built" the space, arranging the necessary relations of the space and the emotional work needed to complete this project. Her grandmother sat at the construction site, providing food and tea for construction workers, telling them to "get this wall to end here or leave a place for windows here ... she was the boss."[21] It was, at first, difficult as a classically trained architect to understand her grandmother's role in the building. It was only when she returned to her family as a part of the family, as relationally centric, that she understood the importance of the emotional and supportive in architecture.

> Neither my academic research in the village site nor my scholarly mapping were enough to allow recognition of the emotional in the built environment. Instead, I could register these emotional operations in the built environment only after I switched from being a scholar back to being a Nubian woman, when my interview subjects stopped being mere objects of study and again became my kin.[22]

The exploration of her own architecture approach rooted in Nubian thought and her matriarchal lineage. As part of her positionality, she takes a stand for Nubian thought and ways of building centered on care. She takes a stand against large-scale Egyptian dam projects and their displacement of Nubian peoples along the Nile River valley certainly, but also the imposed hierarchy of Egyptian-designed settlements. By doing this, she relates resistance and architecture and alternative ways of thinking about space.

Examples of designers' formation, their growth and learning as people, are rare. Professionals write about being inspired by the landscape but not necessarily about how that changed them personally, possibly because it suggests they were not quite whole or "with it" before the transformational event. Academics do not write about their personal transformation as the design disciplines are still in the thrall of positivism, which suggests that the personal point of view is invalid, or at least, does not apply to others. Personal formation is incompatible with the position of expert or knowledgeable one. We have all already arrived.

PRAXIS: REFLECTING ON YOUR HOME PLACE

If reflexivity means an introspection, an examination of one's own background, beliefs and habits that influence how one designs, then the practice of reflexivity is reflection. The best place to begin is to describe and interrogate our landscape(s) of origin. Since place and people are intertwined, this home place will be a strong influence on who we are and how we design. Consider the exercise as autogeography.[23] Like an autobiography, it is the story of an author's life written by the author, except instead of a narrative, the author writes/draws a place. Clare

Cooper-Marcus developed a slightly different version of this exercise during decades of teaching at the University of California, Berkeley.[24] She had students combine ideas of home/self with new conceptions of drawing and design to understand spatial qualities of their home environment.[25]

1. Home place

The exercise works best if the designer uses both writing and drawing. First, engage with one's home place by free writing for one page on its prominent characteristics. What landscape elements shape the place? What is its structure? Who resided in this place? One culture (be as specific as possible) or multiple cultures? How did people move through it? [Note: free writing entails picking up a pen (or keyboard) and writing without setting the pen down. Write without judgement. Ignore grammar. Get the ideas out. If one starts to digress, consider briefly writing down the digression, before returning to the topic of home.]

Draw a map of the extents of your home place. What do you consider its boundaries? Your home could be at one or several scales: house, street, neighborhood, town. Do not use an aerial photo or an accurate map as a base. Bend the map to what is/was important to you. Compile important landscape characteristics into a raw evocation of the place in a collage, layering elements of the map, the topography, the streets, the setting according to your experience of the place.

As an example of my own place of origin ... I grew up in Los Alamos, a small town in the highlands of northern New Mexico that arose during World War II as the United States designed and built an atomic bomb. Decades after the war, the "Lab" still harbored a secrecy of activity, an inaccessibility to our parents' place of work and a seclusion from the rest of (mostly Hispanic and native) New Mexico. Its fences and walls, guard towers channeled movement off and on the hill. Driving downtown, I would see a convoy of military vehicles with manned, machine-gun turrets on the top. They were transporting small blocks of uranium from the airport to the laboratory. Despite the heaviness and militarism of the place, there was an openness to views towards Santa Fe and the landscape of the American West. Watching the cumulonimbus clouds come over the Jemez Mountains on a summer afternoon, I knew I had about half an hour before thunder and lightning would shake the landscape, accentuating my insignificance.

8.9 Collage of the landscape of Los Alamos, focused on the importance of the sky, the topography, and the construction of the atomic bomb that eventually divided the military, the scientific community and the nation.

2. Uniqueness

Based on the written and graphic descriptions of your home place, define what is unique about this place. Why is it unique? Is it because of a different urban form? A different way of moving? A different people relating to their place in a different way? Draw two illustrations of the difference: the first illustration shows what you perceive to be the "standard" or dominant landscape form, the second illustration showing what was unique about your place (if anything). Cross-sections and perspectives work well. For instance, you might look at the street and notice a difference in sidewalk treatment or that there were no streetlights. Alternatively, you might focus on the mix of people in your neighborhood who use outdoor space in a different way.

Note, defining "uniqueness" is challenging. It relies on comparing a home place with what is not unique. The exercise relies on what you consider to be a dominant landscape form. For designers in China, the dominant landscape form may now be the large cities expanding into the country. For designers in Central America, it may be small towns with a commercial center linked by two-lane highways. A good example of a study of uniqueness in a cultural sense is the research of James Rojas into Latino neighborhoods in Los Angeles.[26] As Rojas pursued a master's degree in urban planning, he noticed the planning gospel of the time focused on making room for cars and their movement. He set about documenting the landscape characteristics of streets and yards in the neighborhoods of East Los Angeles and Boyle Heights where he grew up. On these Latino streets, a low wall and gate provide a place for socializing with neighbors, a side yard for family gatherings and wide sidewalks for informal commercial activity. For Rojas, each urban element was unique *in relation to* the dominant urban form in the United States centered around unusable front yards and a suburbanization of the city. By identifying what is unique in a home place, the designer can frame their own approach to design and potentially increase the number of positive attributes of uniqueness and transform dominant practices.

If one word could describe the uniqueness of Los Alamos it is "separate" – the lab from the town, a conclave of white midwestern engineers from the Latinos and Pueblo Indians below, the guard houses, fences and razor wire encircling the places of work. Growing up, we could attend physics lectures on quarks and quasars in the high school auditorium or, like any small town, teenage drinking parties in the surrounding hills. Topography reinforced separation, a change in elevation so that every semi-flat space contained a house or park or business. The Lab occupied two mesas – flat-topped fingers of volcanic material bound on all sides by steep escarpments. To the north, the town sprawled over three mesas: Central, North and Barranca. The town felt precarious; houses perched on cliffs, streets winding up and down hills, the mountains looming over the edge of town.

Despite its separation, the landscape of the lab connects or is in relation to hundreds of other labs, nuclear sites and waste disposal areas throughout the United States. The production of "peaceful" nuclear energy now results in the most poisonous material on earth as a byproduct. Spent uranium fuel is a radioactive material requiring its own depopulated, inert landscape to house and store in deep geological repositories in ways that must discourage people living thousands of years from now from interacting with the place. The byproduct plutonium-239 has a half-life of 24,000 years. Although the Waste Isolation Pilot Plant is 300 miles away from the lab, this landscape is also the landscape of Los Alamos.

Later in college, as I questioned the militarism of Los Alamos, its foundation in the perpetration of unimaginable destruction, I did so in the context of relations with specific families who led the national laboratory – people who I knew to be kind and thoughtful. This personal kindness on the one hand and the perpetuating of a destructive force capable of destroying the world on the other were held in tension. It resulted in a life-long fascination with how people rationalize their decisions, personally and collectively.

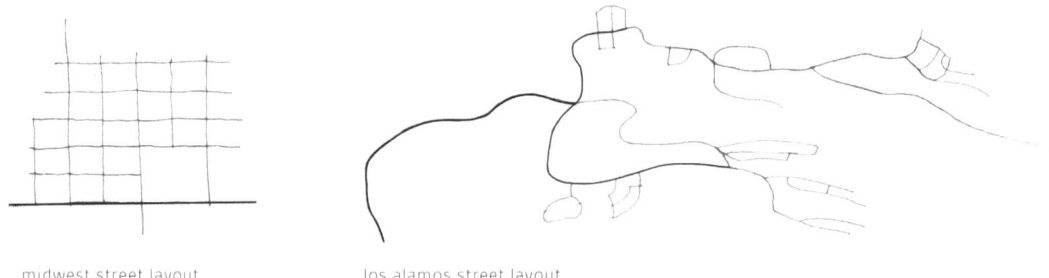

midwest street layout   los alamos street layout

8.10 Contrast between a typical midwestern town (where many lab staff originate) and the street layout of Los Alamos which follows the contours of the mesas.

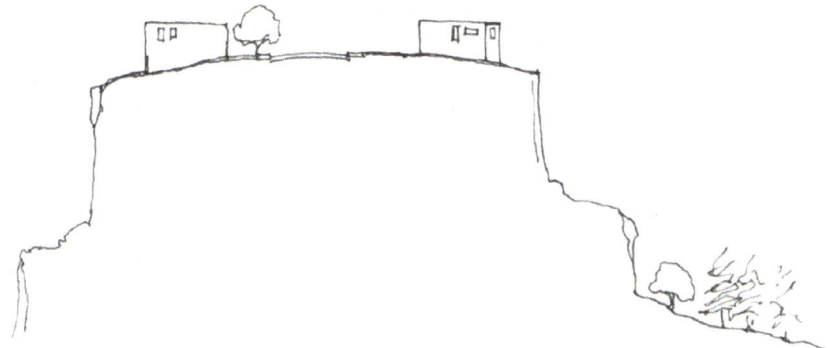

8.11 Cross-section showing the physical forms and shapes that make Los Alamos unique.

3. Design is personal

Look back on your past designs of landscapes (either professional or in studio), what do they have in common? A certain style? An idea of how people might use the landscape? An emphasis on certain landscape elements? Given those prior designs, now apply your autogeography to the way that you design. What is important to you (i.e. gathering places, ecological connections, inclusion of different peoples, clean lines or strong forms, mystery, or practical meeting of needs)?

How does your home place influence your idea of landscape? This could be either a rejection or an embrace of your home landscape. Positive or negative. How does that idea of landscape then influence your evaluation and design of other landscapes? For instance, if you had a "womb" experience of home, growing up, are you trying to re-create this comfortable idea of home in other places? Alternatively, if you moved a lot and are having trouble picking one place, what does it mean to have a series of homes regarding the landscape?

Finally, write a statement on your home landscape (two or three sentences). And write a statement on what a landscape should be (two or three sentences). Highlight what you have learned about yourself and your design process. Acknowledge your tendencies and design approach.

Growing up in Los Alamos influenced the choice to spend my initial years in the profession in the subfields of landscape ecology and restoration, first at the Natural Resources Conservation Service and then at a private wetland ecology firm. The decisions were not a personal response to "landscape guilt," since I had no effect on what happened there as a teenager, but it does suggest an interest in rejecting destructive uses of the landscape, such

8.12 A *parti* of the design for Paynes Prairie's visitor experience.

as extraction, processing, and conversion into fissible substances. Knitting the land back together.

NOTES

1. Dr Haunani-Kay Trask, *From a Native Daughter: Colonialism and Sovereignty in Hawaii*, 2nd edition (Honolulu: Latitude 20, 1999); as quoted in Annabelle Williams, "Haunani-Kay Trask, Champion of Native Rights in Hawaii, Dies at 71," *The New York Times*, July 10, 2021, sec. U.S., https://www.nytimes.com/2021/07/09/us/haunani-kay-trask-dead.html.

2. Trask, *From a Native Daughter*; as quoted in Williams, "Haunani-Kay Trask, Champion of Native Rights in Hawaii, Dies at 71."

3. M.M. Bakhtin, *The Dialogic Imagination: Four Essays* (Austin, TX: University of Texas Press, 2010).

4. Rob Shields, "Meeting or Mis-Meeting? The Dialogical Challenge to Verstehen," *The British Journal of Sociology* 47, no. 2 (1996): 275–294, https://doi.org/10.2307/591727.

5. M.M. Bakhtin, *Speech Genres and Other Late Essays* (Austin, TX: University of Texas Press, 2010), 6–7; as quoted in Robert Shields, "Meeting or Mis-Meeting? The Dialogical Challenge to Verstehen," *The British Journal of Sociology* 287.

6. D. Soyini Madison, *Critical Ethnography: Methods, Ethics, and Performance* (London: SAGE, 2012).

7. Kim V.L. England, "Getting Personal: Reflexivity, Positionality, and Feminist Research," *The Professional Geographer* 46, no. 1 (February 1, 1994): 80–89, https://doi.org/10.1111/j.0033-0124.1994.00080.x.

8. Craig L. Wilkins, *The Aesthetics of Equity: Notes on Race, Space, Architecture, and Music* (Minneapolis, MN: University of Minnesota Press, 2007); June Manning Thomas, "The Minority-Race Planner in the Quest for a Just City," *Planning Theory* 7, no. 3 (November 1, 2008): 227–247, https://doi.org/10.1177/1473095208094822.

9. C. Katz, "All the World Is Staged: Intellectuals and the Projects of Ethnography," *Environment and Planning D: Society and Space* 10, no. 5 (October 1, 1992): 495–510, https://doi.org/10.1068/d100495.

10. Donald A. Schon, *The Reflective Practitioner: How Professionals Think in Action*, 1st edition (New York, NY: Basic Books, 1984), 21.

11. Schon, *The Reflective Practitioner*.

12. Schon, *The Reflective Practitioner*, 40.

13. Hannah Arendt, *Lectures on Kant's Political Philosophy* (Chicago, IL: University of Chicago Press, 1989).

14. Gillian Rose, *Feminism & Geography: The Limits of Geographical Knowledge* (Minneapolis, MN: University of Minnesota Press, 1993).

15. Robert Mulholland, *Shaped by the Word: The Power of Scripture in Spiritual Formation* (Nashville, TN: Upper Room, 1985).

16. Mulholland, *Shaped by the Word*, 94/171 Adobe Digital Edition.

17. Don Schwager, "Formational versus Informational Reading of the Scriptures," *Living Bulwark* (blog), 2019, http://www.swordofthespirit.net/bulwark/bible-study3.htm.

18. Richard Wagamese, *Medicine Walk* (Toronto: McClelland & Stewart, 2014).

19. Wagamese, *Medicine Walk*, 66.

20. Wagamese, *Medicine Walk*, 67.

21. Menna Agha, "Emotional Capital and Other Ontologies of the Architect," *Architectural Histories* 8, no. 1 (December 18, 2020): 6, https://doi.org/10.5334/ah.381.

22. Agha, "Emotional Capital," 1–2.

23. Alan P. Marcus, "Using 'Autogeography,' Sense of Place and Place-Based Approaches in the Pedagogy of Geographic Thought," *Journal of Geography in Higher Education* 47, no. 1 (October 21, 2021): 71–84, https://doi.org/10.1080/03098265.2021.1991290.

24. Clare Cooper Marcus, "Environmental Autobiography," *Room One Thousand* 2, no. 2 (2014), https://escholarship.org/uc/item/1rr6730h.

25. Clare Cooper Marcus, *House as a Mirror of Self: Exploring the Deeper Meaning of Home* (Berkeley, CA: Nicolas-Hays, Inc., 2006).

26. James Thomas Rojas, "The Enacted Environment – The Creation of 'Place' by Mexicans and Mexican Americans in East Los Angeles" (Thesis, Massachusetts Institute of Technology, 1991), http://dspace.mit.edu/handle/1721.1/13918.

Conclusion

According to Mikhail Bakhtin, there is an element in every dialogue that tentatively reaches out to the Other, always contingent upon the response.[1] Those engaged in dialogue extend themselves, exploring the landscape, not as a precursor to conquest but as a "trying on" of a new place as a coat or a shawl is put on over comfortable clothes – reshaping the body. A willingness to listen and to understand must precede visiting the landscape, precede dialogue. Each praxis, whether landscape immersion, critique or dissent, relies on this malleability in the designer to actualize. The landscape is dynamic. We, as a community of design, must be equally open to change.

As part of a dynamic landscape – the slow creep of continents, the migration of people to the coasts, the wind swirling leaves over pavement – we are neither "out there," separate from ourselves and the landscape, but also not "in here," absorbed in an interior conception. Rather, the landscape is an integral partner in an ongoing conversation. We partner with the earth in dialogue, call and response. We are response-able.

Designers immerse themselves in the environment, listening, attempting to understand and *then speaking* – an active response. Speaking back to the landscape to engage with the voice of place. Response entails responsibility. When we as designers spend time listening in the landscape, how will we respond? What is our responsibility? Iris Marion Young, in her book *Responsibility and Justice*, argues for a positive approach to responsibility, as motivation for present, just actions, not accountability for past misdemeanors.[2] This positive approach to responsibility suits the role of designers, shaping landscape dialogue. A designer is responsible for the community in which she designs spaces.

We are responsible to and for the landscape and our interventions within it. The designer through education and years of training becomes qualified to steward the shaping of space to enhance social equity, economic growth, ecological restoration and/or human inhabitation. According to Young, being responsible for something is life as practiced.

DOI: 10.4324/9781003158943-10

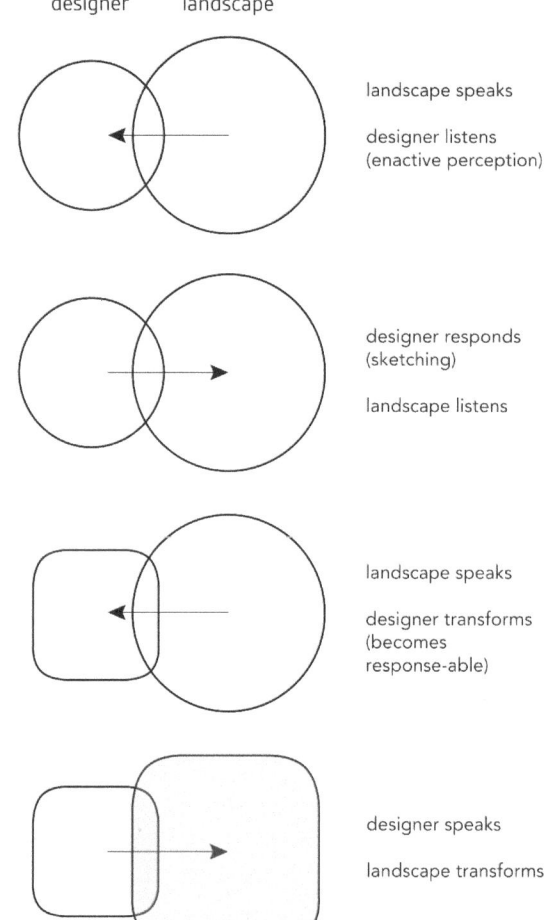

designer   landscape

landscape speaks

designer listens
(enactive perception)

designer responds
(sketching)

landscape listens

landscape speaks

designer transforms
(becomes
response-able)

designer speaks

landscape transforms

9.1 Diagram of an ongoing dialogue between designer and landscape.

What is it we do? Why do we do it? A designer's responsibility *as practiced* begins with space – the needs of a healthy landscape. If we have a responsibility for the landscape and its future, then we must envision that health … what makes for a successfully transformed space? Thus, the myriad social and economic forces acting in space drive the designer to respond. This cannot happen without understanding that which we steward: landscape dynamism and the communities within them – an understanding that is first and foremost an immersive experience.

Do designers have more responsibility than others? That is, should the designer engage with the landscape/community as an "expert" – as a resource of specified knowledge and skills? Young promotes a shared responsibility, acting not as individuals, but as a member of a community (responsibility-with).[3] She draws from Noam Chomsky's 1967 article on the Vietnam War, in which he makes two insights into the existing responsibilities of the expert:[4]

1. The responsibility (for decision-making) is laid more heavily on certain people than others.
2. Experts claim technical knowledge can solve problems, ironically claiming responsibility while shirking it (as a way of distancing themselves personally from the past and present of the place).

The first principle acknowledges the reality of both privilege and experience. Experts practice knowledge – the dissemination of ideas, skills and visions of place – based on education and accumulated experience. The second principle cautions society from relying solely on experts to make decisions, as the expert's temptation, in the face of responsibility, is to deflect to general technical standards – standards which do not consider place. If we act as objective observers juggling categories, poor or inequitable design can be blamed on the information provided, rather than the designer/expert themselves. The responsibility *of* the designer is to respond to landscape challenges using the skills and abilities from a lifetime of practice. To realize our potential as trained individuals in shaping space. To extend our own internal concerns to the exterior world so that people engage with place.

Chomsky's competing claims – that we will always have experts and that these experts simplify the world in order to disclaim responsibility – open the possibility of engagement within a community with a shared, transparent responsibility – a responsibility *with*. A community should not reject the world of technical knowledge in the solving of problems but must practice a proper weaving of the technical into the design process in response to questions about the landscape. Design responsibility embraces the role of designer as someone who listens to a place and only then proceeds to speak. In short … dialogue, but a dialogue that requires time in the landscape to address change, people, health. That requires humility in the face of landscape processes we often do not understand.

DESIGN RESPONSE FRAMEWORK

If design is a speaking response to the landscape, then landscape dialogue encompasses the full spectrum of the traditional design process from site selection to construction documents and oversight. This book focuses

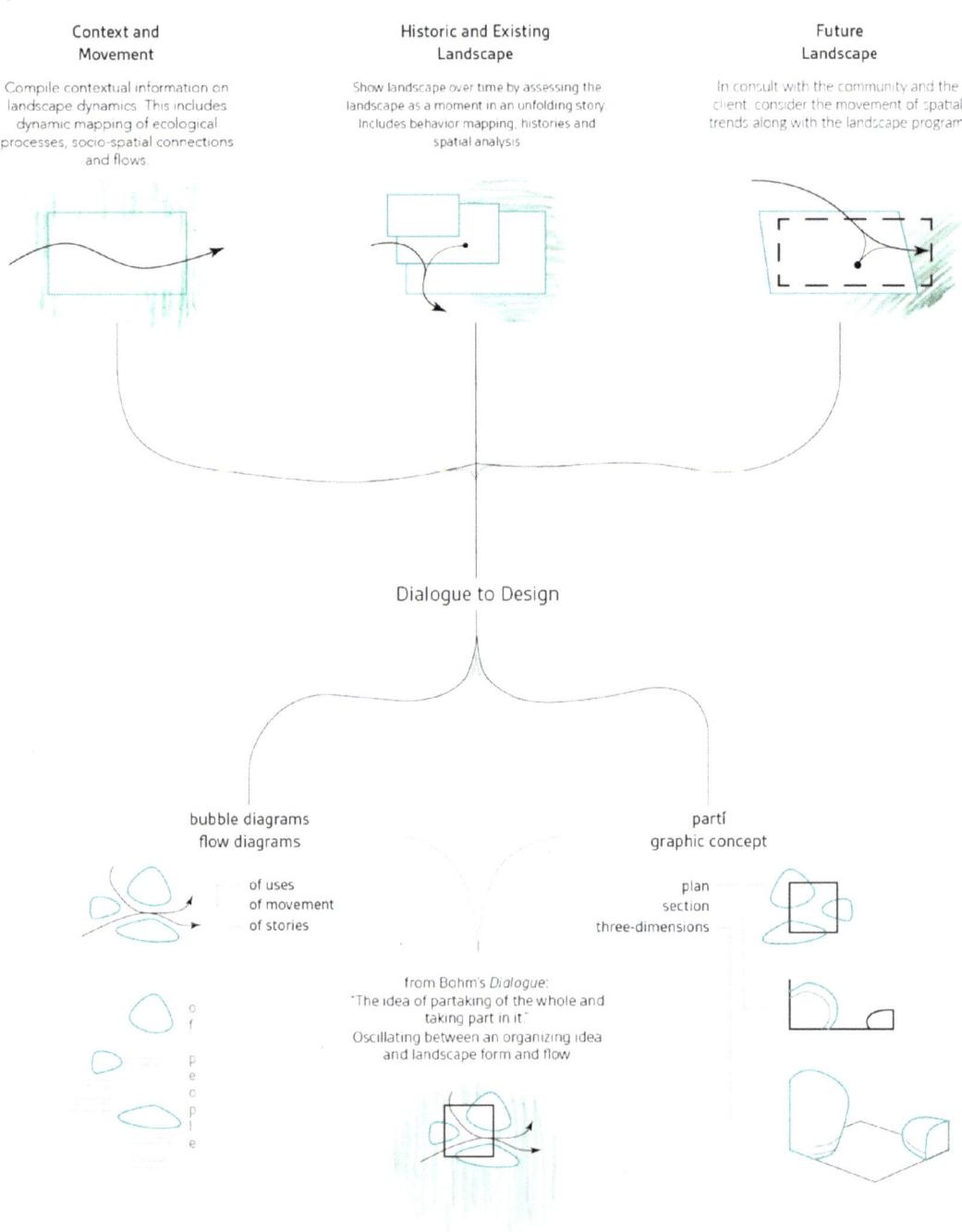

[Design Dialogue Framework]

Context and Movement

Compile contextual information on landscape dynamics. This includes dynamic mapping of ecological processes, socio-spatial connections and flows.

Historic and Existing Landscape

Show landscape over time by assessing the landscape as a moment in an unfolding story. Includes behavior mapping, histories and spatial analysis

Future Landscape

In consult with the community and the client, consider the movement of spatial trends along with the landscape program

Dialogue to Design

bubble diagrams
flow diagrams

of uses
of movement
of stories

of people

partí
graphic concept

plan
section
three-dimensions

from Bohm's *Dialogue*:
"The idea of partaking of the whole and taking part in it."
Oscillating between an organizing idea and landscape form and flow

9.2 Design framework to move from listening to the landscape to speaking with the landscape.

on the early steps in that iterative process. However, it may be helpful to gesture towards later steps in the practice of landscape dialogue. Once a person engages with the landscape, how do they begin the design? To that end, I share a commonly used framework that acknowledges the close connection between dialogue with the landscape and designing with the landscape. Design is also part and practice of dialogue, a response to each landscape's potential.

Throughout the discussion of landscape dialogue, a myriad of practical engagements with place prod the designer to listen and speak with the landscape. This is not a linear process. The designer will have to assess each space on its own terms. However, each assessment of space requires an engagement with its context, its history and its future.

Context and movement

To summarize the practices described in this book, we began by reframing the landscape as a dynamic process, not a static object. Spaces should not be considered in spatial or temporal isolation. Each landscape gets assessed as a changing network of connections to other spaces, to other flows. Context and flow methods discussed include:

- Dynamic mapping (Chapter 1).
- Measuring the soundscape (Chapter 3).
- Immersion (Chapter 4).
- Relational diagrams (Chapter 5).

Historic and existing landscape

If the first category of dialogue assesses a shifting landscape in terms of space, this second category examines the shifting landscape in terms of time, particularly the social and cultural implications of uneven landscape development. Practicing dialogue with a landscape's history and present requires immersion in the space. It requires a critical eye from within. Historic and existing landscape practices include:

- Measuring space with the body and other bodies (Chapter 2).
- Forensic architecture (Chapter 6).
- Trespass and protest (Chapter 7).
- Reflecting on your home place (Chapter 8).

Future landscape

While historic and present analyses of place are important, they are not the ultimate goals of landscape dialogue. Design is a forward-looking process of uncovering opportunities and realizing a vision. (The exception is the setting aside of specific places for remembrance, e.g. a national monument, but even this becomes problematic when a specific people group is honored at the expense of others.) The future landscape is structured in the design process through development of a program, often by the client or public. What should this place become? Assessing the future landscape, particularly a community's vision of a place, is beyond the scope of this book.[5] Landscape dialogue situates this discussion in what this place has been, what this place is saying now, and what would move toward a more just and inclusive space where dialogue continues.

Dialogue to design

At this point in the design process, the designer has listened to the landscape, talked to people and assessed the visions of the client/ community. Certainly for students, but many design professionals as well, the moment of putting pen to paper is fraught with fear and ambiguity. How should one start?[6] I want to acknowledge the real trepidation some feel when they begin the creative process. Akin to writer's block, generating initial sketches can be overwhelming. This is normal, particularly when handed a pen, surrounded by colleagues or fellow students. It helps to think of design as an ongoing conversation as described in this book, as a long-term commitment to another, a marriage rather than a one-night stand. As people speak, conflict, reconcile, so the design speaks, apologizes, goes back and changes into something else. The first effort on trace paper will be thrown away. It will be forgotten. In a similar process to free-writing, the designer is freed from the analytical; criticism recedes and lines appear. Consider the following recommendations:

- Draw on trace (not the computer). At this point, the computer is too slow. Too divorced from the body and its engagement with the landscape.
- Use big markers. Draw coarse lines.
- Be sloppy. You will be more likely to move past these initial efforts if you avoid making them precious. Even the most accomplished designers' first efforts use the vocabulary of the scribble. (See Chapter 5 of Piedmont-Palladino's *How Drawings Work* for a fascinating account of I.M. Pei's drawing of the East Building of the National Gallery of Art).[7]

As one begins design, do not leave dialogue behind. This *is* the dialogue. The challenge is to embrace a landscape vision holistically. That is, while the landscape remains shifting and complex, the design encompasses this complexity in a simple, organizing idea. This takes practice. It takes abstraction. The two most helpful drawings at this stage are the *parti* and the bubble diagram. One related to form, the other to function. Draw multiple *partis* and sketch multiple bubble diagrams. Drawing as a tool of dialogue.

AN ONGOING DIALOGUE

How then to be responsible to, for and with the landscape, not as an expert imposing standards but as a designer engaged with place? I have outlined an approach, rooted in dialogue, that would do just that: embrace the holistic immersion and assessment of the landscape to yield more sympathetic design. It has been difficult to explain in a step-by-step manner, but that is the point. It must be experienced.

Dialogue is not complicated; it is as simple as a conversation. It can be formal or informal, dramatic or calm, focused or wide-ranging. It is often political, in dealing with power and the changing of space to benefit a group of people (and leaving others out). It has its own hierarchies. As designers we can impose on the landscape our own ideas and ignore the words of a space. But soliloquys are uninteresting. As Paulo Freire advocates:

> True dialogue cannot exist unless the dialoguers engage in critical thinking – thinking which discerns an indivisible solidarity between the world and men [*sic*] and admits no dichotomy between them – thinking which perceives reality as process, as transformation, rather than as a static entity – thinking which does not separate itself from action, but constantly immerses itself in temporality without fear of the risks involved.[8]

Practice an ongoing conversation with the landscape. Dialogue is an action. Observe cracks in a pavement, the slumping of soils, the movement of pedestrians along the edge of a space. Then sketch, map, discuss.

Immerse yourself in the landscape. Dialogue requires presence. In absentia, it can only be haphazard and stunted.

Argue with the landscape. It rarely fits into ascribed categories. In conversations with any depth, complete agreement is rare. Empathize with others, particularly those who frequent a space. Creatively engage

9.3 Landscape architecture students in dialogue with a riparian landscape.

others through walking tours, meetings in the landscape and behavioral mapping.

Uncover local histories. Challenge assumptions that affect public space. Make these visible through the telling of landscape stories, whether they are eco-revelatory, e.g. the moving of stormwater from pipes to surface swales, or they coalesce in dissent, changing our ideas of landscape.

Put your whole body into it, your whole self. March in a protest. Measure your stride and then pace a site, not to gather information but to ascribe the site onto your senses.

Place impediments to movement to see how people respond. Then design something to continue the dialogue, to promote dialogue.

Landscape dialogue requires the placement of the body into space as a vulnerable subject. The body "tunes" itself to place, susceptible to the rhythms (or the static) of the landscape as a radio dial ranging over frequencies. The designer's receptivity to a place is not passive but active, combining landscape sensitivity with knowledgeable interpretation of a place's systemic processes. Both sensibility and knowledge reshape the designer as well as the landscape. Our formation leads to its transformation.

NOTES

1. M.M. Bakhtin, *The Dialogic Imagination: Four Essays* (Austin, TX: University of Texas Press, 2010).

2. Iris Marion Young, *Responsibility for Justice* (New York, NY: Oxford University Press, 2011).

3. Young, *Responsibility for Justice*, Chapter 4.

4. Noam Chomsky, "The Responsibility of Intellectuals." *The New York Review of Books*, February 23, 1967.

5. See David de la Pena, L. Lawson, D. Allen, R. Hester, J. Hou and M. McNally (eds.), *Design as Democracy: Techniques for Collective Creativity*, 3rd edition (Washington, DC: Island Press, 2017); Sasha Costanza-Chock, *Design Justice: Community-Led Practices to Build the Worlds We Need* (Cambridge, MA: The MIT Press, 2020), https://library.oapen.org/handle/20.500.12657/43542.

6. See Iain Robertson, *Cultivating Creativity* (New York, NY: New Village Press, 2022).

7. Susan Piedmont-Palladino, *How Drawings Work: A User-friendly Theory* (New York, NY: Routledge, 2018).

8. Paulo Freire, *Pedagogy of the Oppressed*, translated by Myra B. Ramos (New York, NY: Continuum Publishing, 1997), 80–81.

Acknowledgements

I would like to express a heartfelt thank you to the following …

The people of Jones & Jones Architects and Landscape Architects, including Johnpaul Jones, Grant Jones, Ilze Jones and Mario Campos, as well as Paul Olson and Charlie Scott. This book is about an approach to the landscape developed there. It proved messy, difficult to communicate and not always commercially viable, but directly addressed the critical things about landscape and inhabiting a place. Huge thanks to some of the clients willing to embrace this approach, particularly the Confederated Salish and Kootenai Tribes, Gainesville Regional Utilities and the City of Seaside.

My various mentors and colleagues at the University of California, Davis. In particular, Patsy Eubanks Owens, my advisor and research partner who is full of encouragement; Rob Thayer, who started me down this path a long time ago; Susan Handy, for her thoughtful support; and Michael Rios, for his constructive criticism and positivity.

Zannah Matson, Menna Agha and Karen Kubey, my fellow spatial justice fellows at the University of Oregon, where the idea for the book first took hold.

My students at the University of California, Davis who expressed an interest in these ideas and a willingness to try them out in the field.

Unhoused people who have participated in my research. To me, every good analysis of landscape hinges on whether the designer listens to the disadvantaged. Otherwise, we are maintaining a spatial status quo and not truly designing. For the many, many people experiencing homelessness who were so willing to talk, thank you.

My friends, Eric and Jason, for their unwavering, emotional support. And my family, Kathy, Elise and Wren, for the love, advocacy and patience they practice as I wander.

Illustration credits

CHAPTER 1

1.1   Author.
1.2   Author (map based on information from City of Sendai).
1.3   Author.
1.4   Author.
1.5   T. Kishimoto. CC-BY-SA-4.0 International.
1.6   Author.
1.7   Public Domain, Artist Unknown.
1.8   *Netherlands, Voordelta*: NASA Photography.
      The Bursting of St. Anthony's Dike, 5 March 1651: Pieter Nolpe, Metropolitan Museum of Art.
1.9   NOAA.
1.10 CoastView Science.
1.11 Jones & Jones, CAP.
1.12 Oregon Department of Geology and Mineral Industries.
1.13 *1961 Aerial photo of Seaside, Oregon*: Oregon State Archives.
      Tsin-is-tum (Jennie Michel), as she appeared around the turn of the 20th century: Oregon Historical Society. OrHi no 26139. Public Domain.
      Shoreline from 1930 to 2000s: Adapted from Venturato, 2005: NOAA.
      Meeting the train at Seaside, Oregon, 1912. Oregon Historical Society: Kiser Photo Co. Oregon Historical Society.
1.14 Jones & Jones.
1.15 Jones & Jones.
1.16 Leyk et al. 2020, American Association of the Advancement of Science, CC 4.0.
1.17 Venturato, 2005; National Oceanic and Atmospheric Association.
1.18 Author.

CHAPTER 2

2.1   All photos and maps taken by author, courtesy of Jones & Jones Architects and Landscape Architects.
2.2   Bei Guan, courtesy of Jones & Jones.
2.3   Author.
2.4   Author.
2.5   Google Earth.

2.6 Forest and lake photo, sketchbook photo, Author; Topo map from United States Geographic Survey; Graph of error bars; grapherhelp.goldensoftware.com.

2.7 Yorkshire Archaeological Aerial Mapping.

2.8 NPS; Public Domain.

2.9 Author.

2.10 Author.

CHAPTER 3

3.1 Author.

3.2 Author.

3.3 Author.

3.4 Rene Descartes' De Homine, Amsterdam, Daniel Elzevir, 1677. Public Domain.

3.5 Author.

3.6 Author.

3.7 Author.

3.8 Author; Stourhead from Nick Fewings via Unsplash; Ghana gold miners, Dame Yinka, CC-BY-SA 4.0; Catalan atlas, Abraham Cresques, 1375, Public Domain.

3.9 Author.

3.10 Author.

3.11 Author.

3.12 Author.

3.13 Gainesville Regional Utilities, Rheinhard Link, CC-BY-SA 2.0.

3.14 Jones & Jones.

3.15 Jones & Jones.

3.16 Gainesville Regional Utilities, Florida.

3.17 National Park Service, Public Domain.

CHAPTER 4

4.1 Public Domain; California Historical Society, USC Libraries

4.2 Author; Plans by Jones & Jones; Gorilla, ambquinn, Pixabay; Map of Congo, Public Domain.

4.3 Author.

4.4 Bill Cameron, 2007; CC BY-SA 3.0.

CHAPTER 5

CHAPTER 6

6.4   City of Albuquerque.
6.5   Author; Acoma Pueblo from Hasselblad500CM, CC-BY-SA 4.0; Statue and protest by Ali Nuredini, Unsplash; El Camino Real map by National Park Service.
6.6   Kings Cross 1970-71 by Rennie Ellis; Ghanian woman in protest, Fquasie, CC-BY-SA 4.0; Girl walking, Author.
6.7   Joe Mabel, CC-BY-SA 3.0.
6.8   United Nations.
6.9   Travel 2 Palestine CC-BY 2.0 generic.
6.10 Clare Tallamy, Unsplash.
6.11 Author.
6.12 Tapatio, CC-BY-SA 3.0 Unported, 2010.
6.13 Author.

CHAPTER 7

7.1   Author; Police arrest segregationist, Unknown, Public domain; Map of Selma, author; Civil rights march, Colin Lloyd, Unsplash; Edmund Bridge protest photos, U.S Government.
7.2   Author.
7.3   National Park Service.
7.4   Jones & Jones.
7.5   Axel Houmadi, Unsplash.
7.6   JTabor5466, CC BY-SA 4.0.
7.7   *People protesting on a street*: Josh Hild, Pexels.
       Woman in water: Khoa Võ, Pexels.
7.8   Author; Drawing by Le Corbusier (1924), Urbanisme, p. 232; Holland Tunnel by Blue Peep, CC BY-SA-4.0.
7.9   Barak Su, CC BY-SA 4.0.
7.10 Gargarapalvin, CC SA-4.0 International and Google Earth.
7.11 Yükleyenin kendi çalışması, CC BY-SA 3.0.
7.12 Author.
7.13 Clara Ren and Gao Xinyi, University of California, Davis.

CHAPTER 8

8.1   Jones & Jones.
8.2   Jones & Jones.
8.3   Pexel Free Photos.

Index